悦阅
YUEYUE

跟我学做
五星大厨

[澳] 海蒂（Heidi）著　　陆韵菲 译

浙江教育出版社·杭州

Heidi Dugan
小档案

用富有创造力和活力来形容Heidi还远远不够。她的成就来自于她决定有一番作为。作为一个居住在中国的澳大利亚人，她一直是广受欢迎的电视节目《You Are The Chef》的宠儿，同时也是经过认证的健康教练，是跨国企业的咨询师和培训师，负责培训如何进行商务谈判、目标设定、交流与沟通等，帮助中西方高层管理人员更好地了解全球环境和外来工作文化。

Heidi拥有600多万粉丝，被评为"最佳外语电视主持人"和"上海外语频道最受欢迎的主持人"。她曾接受《时尚芭莎》《上海电视》杂志和《上海日报》的采访，也是很多电视节目如《超级家庭》《疯狂的冰箱》《世界爸妈说》的嘉宾。

Heidi旗下的品牌有"Heidi中洋生活"和"Heidi跟我学"。在2016年，Heidi以合伙人的身份加入了料理妈妈。料理妈妈是一家多媒体市场营销公司，致力于帮助和引导母亲和孩子们过上更健康、更快乐的生活。

1996年以来，Heidi一直居住在中国，她嫁给了一位中国功夫和太极大师——张懿，有两个孩子。

Creativity and a dynamic personality don't do justice to Heidi. She's as accomplished as she is determined to make a difference. An Australian living in China, she has been a media darling with the long running popular TV program *You Are The Chef,* a certified wellness coach, a consultant and trainer to multinational companies on negotiation, goal setting, communication and helping both Western and Chinese senior level managers better understand the global environment and foreign work culture.

Heidi has a fan based of over 6 million viewers and has been awarded Best Foreign TV Host and No. 1 Favorite Host on ICS (International Channel Shanghai). She has been interviewed by magazines such as *Harper's Bazaar, Shanghai TV weekly, Shanghai Daily* and guest to numerous TV shows such as *Super Family, Crazy Fridge* and *Shijie Ba Ma Shuo.*

Under Heidi's brands are "Heidi中洋生活" and "Heidi跟我学". In 2016 Heidi joined the Chef Mama team as a co-owner. It is a multimedia marketing company focused on helping and coaching mothers and children on how to lead healthier, happier lives.

Heidi has lived in China since 1996, is married to a Chinese Kung Fu and Taichi Master Zhang Yi, and has two children.

You can connect with me anytime.

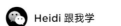 Heidi 跟我学   Heidi 中洋生活

INTRODUCTION
自序

过去17年里，我一直参与《You Are the Chef》的节目拍摄，许多了不起的厨师教会了我烹饪美味佳肴的意义以及如何成为一名出色的厨师。

这本书记录了我在8天的旅程中，采访了20多名厨师并和他们一起烹饪的过程。我了解到，虽然每个厨师都有自己的风格，但是有一些基本的素质和技能，使他们成为和我一起合作过的最棒的厨师。我很高兴能带你一起踏上这个旅程，这样我们可以共同学习如何成为出色的家庭厨师。

烧好菜并不难，难的是如何成为一名好厨师。本书中的食谱并不意味着是简单的菜肴，需要使用各种各样的技术，并需要你的专注才能做好。但是，你会惊讶地发现，通过这个过程你将学到很多东西。为了更好地帮助你，我创建了一个微信公众号"Heidi跟我学"。我拍摄了厨师们如何制作菜肴等的视频和照片，展示了一些小贴士和技巧，涵盖了切工、烹饪以及装盘各方面的技能。你可以登录微信，关注我们，并继续向城市中最好的厨师学习。

在本书的一些页面中，你还将看到鸡尾酒和其他饮料的照片。虽然这本书里没有介绍制作饮料的食谱，但是在我的微信公众号"Heidi跟我学"上

I've been filming *You Are the Chef* for the last 17 years and have been taught by some amazing chefs what it means to cook a great meal, and what it takes to be a great chef.

This book was created over an 8-day journey where I interviewed and cooked with over 20 chefs. I learnt that though each chef has his own style, there are basic qualities and skills that make them some of the best chefs I've worked with. I'm excited to take you along on this journey so we can learn together what it takes for us to become great home chefs.

To cook a good dish is not difficult, to be a good chef is. The recipes in this book are not meant to be basic dishes; they use a variety of techniques and will require you to focus to be able to cook them well. But you will be amazed at how much you'll learn through the process. To help you I have created a Wechat account "Heidi跟我学". I have filmed the chefs making the dishes and many more. There are videos and photos which show the tips and tricks, techniques and skills, of everything from chopping to cooking and plating. You can log on, follow me and continue to learn from the best chefs in town.

Throughout the pages of this book you will also see images of cocktails and other beverages. Although the recipes are not in this book, I do have them on "Heidi跟我学" Wechat account, along with some of the best bartenders and mixologists in town.

Corelle Dinnerware
康宁餐具

有这些食谱，还有一些城市中最好的酒吧调酒师的介绍。

一定要在家里尝试制作饮料和菜肴，然后到餐厅去品尝大厨们的厨艺。这是一个比较你烹饪的菜肴和大厨们菜肴的很好的方式，如果你有问题可以问厨师，他们会很乐意指导你。

过去的8天正好提醒了我，我们应该向厨师们致敬。本书中的所有厨师们——每天用心烹饪的，长时间工作到深夜的，周末和节假日也在工作的，无论我们上班还是休息，他们都兢兢业业地在工作，却很少得到感谢的所有厨师们——我想感谢你与我们分享如何成为一名优秀厨师的核心本质。我向你们每一位举杯致意，我从内心感谢你们带给我们美好的体验。

感谢你们。

还有正在读此书的你们，记得下次外出就餐有很好体验的时候，别忘了向厨师们致敬。

Definitely give the drinks and dishes a try at home, after you have, head down to the restaurant to taste them. It's a great way to compare your dish with the chefs, and if you have questions, ask the chef, they'll be happy to guide you.

These past 8 days were a wonderful reminder to me how much we have to be grateful to chefs. So to all the chefs in this book who pour their heart into their work everyday, who work long hours into the night, who work on weekends and festive holidays, who work when we work and when we rest, who work and very rarely get thanked. I want to thank you for sharing with us the core essence of what it means to be a great chef. I raise my glass to each and every one of you, and thank you from my heart for all of the wonderful experiences you have given us.

Thank you.

And for you my friends reading this book, remember next time you dine out and have a good experience, be sure to ask for the chefs and thank them in person.

今天就加入"Heidi 跟我学"大家庭吧！

Best of British Event
《英伦精选》录制活动现场

DEDICATION
题献

首先，我要感谢所有参与此书编写的厨师们。
谢谢你们愿意花时间分享你们这么多年以来积累的经验。

感谢我的家人张懿、张朔燕和张瀚文一如既往地支持我、相信我！

感谢我亲爱的朋友Shannon，特地从澳大利亚飞来设计、拍摄菜肴，
也谢谢她花了大量的时间通过电话来引导我成书的感觉和书的布局。

这本书的出版是我们共同努力的成果。

Firstly, I want to thank all of the chefs who took part in this book.
Thank you for spending the time and sharing what has taken you years to learn.

To my family Zhang Yi, Syana and Oden who support and believe in me.

To my dear friend Shannon who flew from Australia to style and photograph the dishes,
and for the hours spent on the phone guiding me on the feel and layout of the book.

This book is a combination of all of our efforts.

CONTENTS
目录

Heidi Dugan
小档案 ·········· III

INTRODUCTION
自序 ·········· IV

DEDICATION
题献 ·········· IX

Day 1
TOOLS
工具 ·········· 001

Day 2
INGREDIENTS
食材 ·········· 011

Day 3
SKILL
技巧 ·········· 030

Day 4
CLASSICS
经典 ·········· 047

Day 5
HEART
用心 ············ *059*

Day 6
UNIQUE STYLE
独特的风格 ············ *078*

Day 7
EXPERT
成为专家 ············ *099*

Day 8
HAVE FUN
玩得开心 ············ *119*

SPECIAL THANKS
特别感谢 ············ *134*

MEASUREMENTS
OVEN TEMPERATURE
计量单位对照
烤箱温度对照 ············ *135*

Day 1

TOOLS
工具

| BRADLEY TURLEY | STEVEN ER |

一位好厨师有他自己的一套工具，无论是刀、平底锅，还是厨房电器，他都会很爱惜。没有好的工具，在厨房里干活会变得困难。

如果你刚刚开始学习烹饪，我建议你给自己买一套好一点的刀具。它们不一定是市场上最好的，但它们应该是锋利的。至于其他工具，去找一些你能够负担得起的东西。你一开始最好不要买很多东西，但是你买的东西，一定要质量好，避免发生意外。作为初学者，我建议你的工具包里应配备一套刀具，一套大小不同的锅，一台强力料理机和一台厨师机。一旦有了这些工具，烹饪将会变得更加容易，基本上可以做任何你想做的料理。

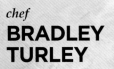

chef
BRADLEY TURLEY

San Francisco, America

旧金山，美国

restaurant
Hai by Goga

Contemporary Asian inspired Californian cuisine

当代亚洲风味的加利福尼亚式佳肴

Black Cod with Miso

味噌黑鳕鱼

2—3 块黑鳕鱼	2–3 black cod fillets
腌酱	Marinade
1/4 杯清酒	1/4 cup sake
1/4 杯料酒	1/4 cup mirin
4 汤匙白色味噌酱	4 tbsp white miso paste
3 汤匙白砂糖	3 tbsp sugar
菠菜酱	Spinach Sauce
1 束菠菜	1 bunch spinach
1 根青葱	1 stem spring onion
1 茶匙橄榄油	1 tsp olive oil
1 汤匙味增酱	1 tbsp miso paste
盐和胡椒粉	salt & pepper
装盘	Plating
6 个香菇，切薄	6 shiitake mushrooms, sliced thinly
6 个樱桃番茄，对半切	6 cherry tomatoes, sliced in half
3 根青葱，只要绿色部分	3 stems spring onions, green section only

【制作鳕鱼】将腌制用的酱料倒在一个带盖的容器里，充分混合，然后放置一旁。留一些，装盘时用。用厨房用纸吸去鱼块上多余的水分。将鱼块放入盛有腌酱的容器中，盖上盖子，放入冰箱冷藏至少 24 小时。

在制作鳕鱼前，先将烤箱预热至 200℃。轻轻拭去鱼块上多余的味噌酱，但千万不要冲洗。将鱼块放在烤架上，两面都用小火烧烤，直到表皮变成棕色，再将鱼块放进烤箱烤 10—15 分钟。

【制作菠菜酱】将水煮沸，加入菠菜焯 30 秒。关火，将水沥干。将菠菜和青葱、味增酱和橄榄油一起放入搅拌机搅拌。加入适量的盐进行调味。

【制作蔬菜】将锅中的橄榄油加热，加入香菇，炒 3—4 分钟；然后加入番茄，再炒 2 分钟；最后加入适量的盐和胡椒粉进行调味。

【装盘】加入几滴腌酱，再将 3—4 茶匙的菠菜酱倒在餐盘的中间。将鳕鱼放在酱上，然后将煮熟的蔬菜放在鳕鱼上，最后再撒上少许切碎的青葱即可享用。

Tips: 如果你没有时间浸泡24小时,你仍然可以做这道菜。即使浸泡了15分钟,它品尝起来还是棒极了!

chef
STEVEN ER

Shanghai, China

上海，中国

restaurant
Henkes

Modern Mediterranean
A gourmet village of authentic
urban dining experiences
in the heart of Shanghai

现代地中海式菜肴
在上海市中心体验地道美食

Braised Lamb Shank with Eggplant, Tomato & Basil

香料烩羊腿

4块羊羔后腿肉（500克）	4 lamb shanks (500g)
50克茄子	50g eggplant
100克番茄	100g tomatoes
10克甜辣椒粉	10g smoked paprika
2克罗勒	2g basil
50克面粉	50g flour
500毫升红葡萄酒	500ml red wine
5克盐	5g salt
5克胡椒粉	5g pepper
2克百里香	2g thyme
10毫升油	10ml oil
200毫升鸡汤	200ml chicken stock

【烤羊腿】烤箱预热至180°C。将盐、胡椒粉、甜辣椒粉和面粉混合均匀，然后将混合好的面粉轻轻地涂抹在羊腿的两面。用中火加热煎锅，倒入橄榄油，放入羊腿，把羊腿两面煎成焦黄色。将羊腿从煎锅中取出，放入一个砂锅中，加入红葡萄酒、百里香和鸡汤。将砂锅放在炉子上加热，煮至沸腾，然后盖紧锅盖，把砂锅放入180°C的烤箱中烤40分钟。当羊肉看上去像要从骨头上掉下来的时候，说明已经烤熟了。

【茄子和番茄】茄子、番茄切粗块，撒上适量的盐和胡椒粉。用橄榄油把茄子煎至金黄，再加上番茄，轻炒。

【装盘】当羊腿酥软时，从砂锅中取出羊腿。将砂锅放在炉子上，不用加盖。调至中火，收汁。品尝一下，根据口感，加入适量的盐和胡椒粉，再加入茄子和番茄。把羊腿放在一个深底的盘子里，淋上调好的蔬菜汁水，最后将新鲜的罗勒放在上面做点缀，即可享用。

这道菜第二天吃起来味道也不错，所以可以多做一些，这样第二天也可以当午饭吃。

Day 2
INGREDIENTS
食材

| EDUARDO VARGAS | ALEXANDER BITTERLING | BRIAN TAN |

无论你去问哪一位厨师，他都会说：每道菜的核心就是好的食材。好的厨师对他们所使用的食材的牌子、新鲜程度都非常讲究。正确选择食材是所有美味佳肴之本。如果发现你处理的肉太硬或者打发的巧克力混合物不够丝滑和细腻，那么问题很有可能出在食材的选择上而不是烹饪技术上。

高品质的食材不一定是贵的，但它们一定是新鲜的。经常检查产品外包装的生产日期，尽量购买距生产日期近的产品。

chef
EDUARDO VARGAS

Lima, Peru
利马，秘鲁

restaurant
Colca

Full flavoured Latin cuisine
全风味拉丁佳肴

Northern Style Duck

北方风格的鸭子

2个白洋葱，切成块状	2 white onions, cut into chunks
18个蒜瓣，保留外皮	18 garlic cloves, skin on
1捆韭菜（仅限白色部分）	1 leek (white part only)
1个中等大小的胡萝卜	1 medium carrot
1个去核的苹果	1 apple, without core
2捆芹菜茎	2 celery stalks
8个杏子对半切	8 apricots cut in half
3个番茄（去皮）	3 tomatoes (no skin)
6个鸭腿	6 duck legs
375毫升白葡萄酒	375ml white wine
1.5升鸡汤	1.5l chicken stock
1.5升牛肉汤	1.5l beef stock
4片新鲜的鼠尾草叶	4 sage leaves
1根百里香	1 sprig of thyme
2捆新鲜意大利欧芹	2 Italian parsley sprigs
3片干的月桂叶	3 bay leaves
盐	salt
黑胡椒粉	black pepper
橄榄油、素油	olive oil, vegetable oil

黑啤"黏稠"饭 / Black Beer "Sticky" Rice

500克泰国长粒米，洗净	500g Thai long grain rice washed
6瓣蒜，浸入橄榄油和酱中	6 garlic cloves, in olive oil & pureed
1个红洋葱切丝或切丁	1 red onion shredded or chopped finely
2个去皮番茄，切丁	2 tomatoes, no skin & chopped finely
1个皮奎洛甜红椒，切末	1 piquillo pepper chopped finely
1/2个胡萝卜去皮，切小块	1/2 carrot peeled & cut finely
250毫升黑啤酒	250ml black beer
1杯混合了鸡汤的香菜泥	1 cup of cilantro leaves blended with chicken stock
350毫升鸡汤	350ml chicken stock
盐、黑胡椒粉、白砂糖	salt, black pepper, white sugar
橄榄油、素油	olive oil, vegetable oil

将烤箱预热至160°C。

将洋葱、大蒜、韭菜、胡萝卜和芹菜茎切成1厘米大小的丁状。将素油倒入锅里后加热，然后放入鼠尾草、百里香、欧芹和月桂叶。加入切丁的蔬菜，煮至变黄、变软。再加入杏子、苹果和番茄，煮1分钟。加入适量的盐和黑胡椒粉调味，然后将蔬菜混合物装入碗中，放置在一旁。在鸭腿上划几刀对角线，然后用适量的盐和黑胡椒粉调味。将可以放进烤箱烘烤的托盘低温加热，倒入适量的橄榄油，将鸭腿烤至金黄。加入蔬菜混合物，再倒入白葡萄酒。小火烹煮直到酒精挥发。

在另一个碗中将鸡汤和牛肉汤混合，然后将1/3的混合汤倒在鸭腿和蔬菜上。小火加热，煮至汤沸腾。关火，用锡纸覆盖住托盘上的食材，放进160°C的烤箱烘烤3小时。每隔1小时就将剩余的高汤倒在鸭腿上，以确保烹饪的过程中鸭腿不会太干。

从烤箱中取出托盘，放置在一旁让它冷却。将鸭腿小心翼翼地放进另一个容器里。将汤汁倒在鸭腿上，去除蔬菜。当鸭腿冷却后，去除皮下脂肪。

秘制红辣椒酱

12 个蒜瓣，去皮
20 个中等大小的红辣椒
250 毫升特级初榨橄榄油

特制的可利欧莎莎酱

1 个番茄去皮，切丁
1.5 个中等大小的白洋葱，洗净，切丁
1 个新鲜的黄椒，去籽，洗净，切丁
1 个中等大小的青椒，去籽，洗净，切丁
2 汤匙香菜叶子，洗净，晒干，切碎，
3 个青柠榨成汁
半茶匙秘制红辣椒酱
4 茶匙特级初榨橄榄油
盐

加入餐盘

3/4 杯熟玉米粒
3/4 杯新鲜煮熟的青豆
1 个辣椒，切末
2 茶匙香菜，切碎
350 毫升鸡汤
1 个青柠榨成汁
特级初榨橄榄油

Special Red Chili Aioli

12 garlic cloves peeled
20 medium red chilies
250ml extra virgin olive oil

Special "Criolla" Salsa

1 tomato de-skinned and cut finely
1.5 white onion medium, washed, diced
1 yellow chili, de-seed, washed, diced
1 green chili medium, de-seed, washed, diced.
2 tbsp cilantro leaves, washed, dried, cut chopped
the juice of 3 limes
1/2 tsp specail red chili aioli
4 tsp extra virgin olive oil
salt

Plating

3/4 cup cooked corn kernel
3/4 cup cooked fresh green peas
1 chili chopped
2 tsp cilantro chopped
350ml chicken stock
the juice of 1 lime
extra virgin olive oil

【黑啤"黏稠"饭】将米洗净、沥干。在锅里倒入一些橄榄油和素油，加热，加入洋葱炒2分钟。加入大蒜、番茄和皮奎洛甜红椒，烹煮2分钟直到混合物变软。加入胡萝卜，烹煮1分钟后，再加入香菜泥，烹煮2分钟。用盐和1/2茶匙的白砂糖进行调味。加入大米继续烹煮，不停地搅拌大米直到变热。大米变热后，加入黑啤酒，煮到酒精挥发，然后加入足够淹没大米的鸡汤。用盐进行调味，混合后用锡纸覆盖住，放进烤箱烘烤至米饭变熟。

【秘制红辣椒酱】将辣椒洗净、去籽。将蒜瓣底部的蒂去除。将辣椒、蒜瓣和大量水倒入锅中，用小火加热30分钟，把水倒掉，再重新加水，把刚才的过程重复一遍，继续用小火加热30分钟。蒜瓣和辣椒变软后，倒入特级初榨橄榄油搅拌。

【特制的可利欧莎莎酱】在一个碗里加入洋葱、番茄、黄椒和青椒，并用盐调味，再加入青柠汁、秘制红辣椒酱和橄榄油，充分拌匀，最后加入切碎的香菜叶子。

【加入餐盘】将鸭腿放在一个小的平底锅里，加入现有的汤汁，用小火温热。然后小心翼翼地取出鸭腿，将鸭小腿和大腿分开，去骨，切粗块。在平底锅中倒入一些橄榄油，将大腿有皮的那面煎至酥脆。往锅里倒入鸡汤以及烹煮鸭的时候流出的汤汁，收汁（汤剩一半）。加入玉米粒、青豆、切碎的鸭肉和啤酒米饭，拌匀，直到混合物变得黏稠。关火，加入切碎的香菜末，滴一滴橄榄油和青柠汁。轻轻拌匀，与酥脆的鸭腿和特制的可利欧莎莎酱一起享用。

Follow
Heidi 跟我学

您将会获得

> 时令食谱
> 特殊节假日的甜美食谱
> 启发性的小贴士
> 工作日晚餐的灵感

今天就加入
"Heidi 跟我学"
大家庭吧!

restaurant

Thought for Food

Pure flavours with innovative textures

创新纯风味佳肴

chef
ALEXANDER BITTERLING

Brandenburg, Germany

勃兰登堡，德国

Duck Terrine

砂锅烧鸭

砂锅烧鸭	Duck Terrine
415 克鸭胸肉	415g duck breast
190 克鸡肝	190g chicken liver
85 克青葱	85g shallot
16 克新鲜生姜	16g fresh ginger
10 克大蒜	10g garlic
1.8 克香菜籽	1.8g coriander seed
4 片月桂树叶	4 bay leaves
4 克黑胡椒粉	4g black pepper
4 克迷迭香	4g rosemary
17 克红糖	17g brown sugar
15 克盐	15g salt
240 毫升水	240ml water
5.8 克食用琼脂	5.8g agar

樱桃&四川辣椒酱	Cherry & Sichuan Pepper Sauce
125 克樱桃	125g cherries
15 克红糖	15g brown sugar
70 克柠檬汁	70g lemon juice
25 克苹果醋	25g cider vinegar
7 克四川辣椒,切碎	7g sichuan pepper crushed

装盘	Plating
3 颗开心果,切碎	3 pistachios, chopped
豆芽菜	beansprouts
无花果和西梅(可供选择)	fig and plum (optional)

【砂锅烧鸭】烤箱预热至120°C。在锅中倒入水,加入食用琼脂、大蒜、月桂树叶、迷迭香、黑胡椒粉、香菜籽、青葱和生姜。将水煮沸,均匀搅拌,直至食用琼脂煮熟,关火。将鸡肝在水中冲洗干净,拍干。剥去鸭胸上的皮。将鸭胸肉以及鸡肝切碎。把切好的肉三分之一放在一边,剩余的放入搅拌机,加入适量的盐,高速搅拌1分钟。将锅里的食材倒入搅拌机和肉一起搅拌30秒,再加入剩余的鸭胸肉和鸡肝,均匀地搅拌。
准备一个扁平的模具,用硅油纸覆盖模具底部。往模具中倒入搅拌好的肉末,用盖子盖上,放进预热过的烤箱中烤45分钟。冷却后,可以放进冰箱冷藏4小时。

【樱桃&四川辣椒酱】洗净樱桃并去核,切丁。将柠檬汁、烘烤后压碎的四川辣椒、苹果醋和红糖混合在一起,倒在樱桃上,真空密封。静置1小时后,将调好的汁倒入锅中用中火加热,收汁,直至汁水减少三分之一,然后关掉炉火,使汁水冷却。汁水完全冷却后,和樱桃充分混合均匀。

【装盘】从冰箱中取出烧鸭,在室温中放置10分钟,并切割成长方形(或需要的形状),放在盘子上,再撒上少量切碎的开心果。还可以搭配一些烤好的面包,最好是酵母面包。
在烧鸭上加上一些腌制的樱桃,淋上汁,在最上面也可以加上一些新鲜的豆芽菜和无花果。

chef
BRIAN TAN

Terengganu, Malaysia

登嘉楼，马来西亚

restaurant
hoF

Cocktail and chocolate dessert lounge

以鸡尾酒和巧克力甜点
为特色的休闲餐厅

Chocolate Mousse Soy Skin Millefeuille

巧克力慕斯豆皮酥拿破仑

慕斯

100 克马斯卡彭奶酪
10 克细砂糖
半块融化的明胶
210 克融化的黑巧克力
250 克半打发奶油

焦糖豆皮酥

6 张豆皮
80 克细砂糖
120 克融化的黄油

Mousse

100g Mascarpone cheese
10g fine sugar
1/2 pc melted gelatine
210g melted dark chocolate
250g semi-whipped cream

Soy Caramel Puff Pastry

6 sheets soy skin
80g fine sugar
120g melted butter

【慕斯】将室温下的马斯卡彭奶酪和细砂糖搅拌均匀，作为底料。明胶需要浸泡在一碗冰水中，软化后沥水。将软化的明胶用小火加热，直至融化。然后加入马斯卡彭奶酪底料，一边搅拌，一边加入融化的巧克力。接着，轻轻地拌入奶油，直至混合物细腻有光。切记，请勿过度搅拌，不然会导致混合物分离。最后，将混合物倒入碗中，冷冻。

【焦糖豆皮酥】在大豆皮上均匀地刷上融化的黄油和少量细砂糖，然后铺上另一层大豆皮，再刷黄油和细砂糖，直到铺上最后一层。将上述多层大豆皮切成细长条，然后卷成两个不同大小的大豆环。在180°C的烤箱中大约烘焙15分钟，直至焦糖变成棕色，然后将大豆环从烤箱中取出，让其冷却。接下来在两个不同大小的大豆环中填充巧克力慕斯，之后即可装盘享用。

关注微信公众号"Heidi跟我学"，查看制作这款慕斯的步骤。

Mango Chocolate Ravioli with Osmanthus

芒果巧克力丸子

慕斯	**Mousse**
100 克马斯卡彭奶酪	100g Mascarpone cheese
10 克细砂糖	10g fine sugar
半块融化的明胶	1/2 pc melted gelatine
210 克融化的黑巧克力	210g melted dark chocolate
250 克半打发奶油	250g semi-whipped cream
开心果脆	**Pistachio Brittle**
20 克细砂糖	20g fine sugar
20 克开心果	20g pistachios
装盘	**Assemble & Plating**
1 个熟芒果	1 mango, ripe
30 克法芙娜香脆珍珠	30g Valrhona crunchy pearls
20 毫升桂花酿	20ml osmanthus syrup
新鲜采摘的桂花	fresh osmanthus

【慕斯】将室温下的马斯卡彭奶酪和细砂糖搅拌均匀，作为底料。明胶需要浸泡在一碗冰水中软化，然后沥水。将软化的明胶用小火加热，直至融化。然后加入马斯卡彭奶酪底料，一边搅拌，一边加入融化的巧克力。接着，轻轻地拌入奶油，直至混合物细腻有光。切记，请勿过度搅拌，不然会导致混合物分离。最后，将混合物倒入碗中，冷冻。

【开心果脆】将糖用小火加热至融化，然后加入开心果，不断搅拌，直到焦糖裹附在开心果上。把做好的开心果脆倒在硅油纸上冷却。

【装盘】将芒果切成薄片，放在保鲜膜上。在芒果切片上挤上巧克力慕斯，撒上一些香脆巧克力珍珠和开心果脆，裹成球状，在冷藏室和冷冻室各放置半小时。享用前，小心拨开保鲜膜，把丸子摆在盘中心，涂上桂花酿并撒点鲜桂花装饰。

Day 3
SKILL
技巧

| STEPHAN STILLER | TONY SU |

如何成为米其林星级厨师或者成为米其林星级餐厅的一员？毋庸置疑，这没法一步到位，但至少可以先这样做起：厨房总是一尘不染，事先做好准备工作（这样，厨师就不会手足无措，无论发生什么事，他们总能沉着冷静地面对），敢于尝试新的东西。

顶级厨师一定具备了扎实的基本功，即基本的切工和烹饪技能。为了练好基本功，厨师需要花时间慢慢来学习技巧，然后在这个基础上会衍变出更多创新的技巧。

chef
STEPHAN STILLER

Hamburg, Germany

汉堡，德国

restaurant
Taian Table

Modern innovative European cuisine

现代创新欧式佳肴

Chawan Mushi, Caviar & Broccoli Coulis

茶碗蒸

日式高汤
20 克日本海菜
30 克鲣鱼片
6 个干香菇
1 升水

茶碗蒸
250 毫升日式高汤
20 毫升生抽酱油
8 毫升料酒
3 个蛋

花椰菜酱汁
300 克花椰菜
20 克切碎的葱
550 克水
2 克墨西哥辣椒汁
4 克柠檬汁
盐和胡椒粉

Dashi Stock
20g kombu seaweed
30g bonito flakes
6 dry shiitake mushrooms
1l water

Chawan Mushi
250ml dashi stock
20ml light soy sauce
8ml mirin
3 eggs

Broccoli Coulis
300g broccoli florets
20g sliced shallot
550g water
2g tabasco
4g lemon juice
salt & pepper

【日式高汤】把所有食材放进一个汤锅，小火炖（约80°C，不要煮沸）30分钟。过筛后让汤汁冷却。

【茶碗蒸】用搅拌机将所有食材粉碎，然后过滤，让液体在室温下静置约1小时，防止产生气泡。接着将液体倒入单独的碗碟中，并用保鲜膜盖好，置于85°C下，蒸35—45分钟，直到液体凝固。

【花椰菜酱汁】在中号平底锅中倒入橄榄油，把青葱炒香。加入花椰菜，炒1—2分钟后加水，用高火煮3—4分钟。冷藏后，将混合物打成酱汁，用适量的盐、胡椒粉、墨西哥辣椒汁和柠檬汁调味。过筛，把酱汁放置在一旁备用。

【装盘】在茶碗蒸顶部淋上一层花椰菜酱汁和橄榄油，撒上胡椒粉，并在它的中心加一匙鱼子酱，即可享用。

Umeboshi Cream Cheese

梅子奶油奶酪

250 克鸡蛋	250g eggs
170 克糖	170g sugar
170 克切块的黄油	170g diced butter
500 克切块的奶油奶酪	500g diced cream cheese
1 撮盐	1 pinch of salt
60 克梅子酱	60g umeboshi paste

草莓蛋白霜	**Strawberry Meringue**
3 个蛋白	3 egg whites
180 克糖粉	180g icing sugar
1 杯青柠汁	1 cup lime juice
抹茶粉	matcha powder
冻草莓干	freeze-dry strawberries

烤箱预热至65°C。从冰箱中取出黄油和奶油奶酪，放在室温下软化。打发鸡蛋和糖，直至变稠。加入切块的黄油，然后加入切块的奶油奶酪。用厨师机搅拌3—4分钟，搅成柔滑状。加入梅子酱，搅拌均匀。最后将混合物装入碗中，静置一晚。

【草莓蛋白霜】在蛋清中加入糖，打发成蛋白霜，之后加入青柠汁。将烘焙纸放在烤盘上。将蛋白霜挤在烤盘上，1.5—2毫米厚。撒上抹茶粉和少许冻草莓干。在65°C的烤箱中强力烘干，至少4—5小时，直到蛋白霜变得非常干脆。

【装盘】从冰箱中取出装有奶油奶酪的碗，撒上抹茶粉。再把蛋白霜切碎，将其放入奶油奶酪中。

chef
TONY SU

Shanghai,
China

上海，中国

restaurant
Ming Court
Cordis,
Shanghai, Hongqiao

Contemporary
Cantonese cuisine

当代粤式珍馐佳肴

Stir-fried Chicken, Ginger & Tangerine Peel in Clay Pot

鲜沙姜橙皮煎鸡肉

150 克鸡肉	150g chicken
100 克葱	100g green onion
50 克鲜沙姜	50g freshly ground ginger
10 克橙皮	10g orange zest
10 克香菜	10g caraway
10 克红椒	10g chili pepper
1 茶匙淀粉	1 tsp corn starch
蚝油、鸡粉、糖、老抽、生抽	oyster sauce, chicken powder, sugar, dark soy sauce, light soy sauce

将红椒切细丝，香菜杆切段。用蚝油、鸡粉、糖、鲜沙姜粒腌制鸡肉。

用少许油将鸡肉煎至七成熟倒出。将准备好的葱粒煸炒出味，倒入鸡肉，放入红椒丝、香菜杆、橙皮丝与鲜沙姜一同煸炒，并放入生抽、老抽、鸡粉。

勾芡少许，淋油出锅装盘即可。

做好这道菜的关键是切工。

关注微信公众号"Heidi跟我学"，
看看Tony是如何做到的。

Braised Asparagus & Bamboo Fungus with Crab Meat & Roe

珊瑚映窗纱

150 克竹荪	150g bamboo fungus
150 克芦笋	150g asparagus
75 克蟹肉	75g crab meat
50 克蟹黄	50g crab roe
100 毫升高汤	100ml stock
盐、糖、鸡粉	salt, sugar, chicken power

将芦笋与竹荪用沸水煮一下，将蟹肉与蟹黄用沸水烫至七成熟。

把芦笋串进竹荪里装盘。起锅，高汤中放少许盐、鸡粉、糖，勾芡均匀，淋在竹荪串上。

锅里留少许高汤，放入蟹肉与蟹黄调味，然后捞出，摆放均匀，淋油，浇在竹荪串上即可。

Day 4
CLASSICS
经典

| NAT ALEXANDER | SEAN CLARK |

传统菜肴可以作为学习烹饪的最佳起点，它们通常使用传统、基本的技巧，掌握这些是成为一位优秀厨师的基础。这也是为什么所有优秀的烹饪学校都把传统菜肴作为核心课程的原因。因为一旦你学会了如何正确烹饪传统菜肴，你就可以汲取更多的创意灵感。

无论是中国菜、法国菜还是西班牙菜，它们都有令人惊叹的传统菜肴，这些菜肴将教会你那个菜系的精髓：食材的使用，烹饪前的准备和烹饪的方法。传统的食物就像一双可以看到过去的眼睛，反映了一个国家的文化。当你明白这个道理的时候，你就可以烹饪那个国家的传统菜肴了。

048 / 049

restaurant
Yang Jing Bang

A pop-up concept exploring Chinese food through foreign eyes

一个中西菜肴结合的
快闪概念餐厅

chef
NAT ALEXANDER

Bath,
United Kingdom

巴斯，英国

Tips:

在切猪肉时,先冷冻30分钟,这样就会容易许多。☺

Lurou Cottage Pie

牧羊人派

800 克猪肩肉	800g pork shoulder
2 茶匙油	2 tsp oil
1 个中等大小的洋葱	1 medium onion
1 个中等大小的胡萝卜	1 medium carrot
25 克生姜	25g ginger
1 个八角茴香	1 star anise
1 块桂皮	1 stick of cinnamon
2 片月桂叶	2 bay leaves
2 汤匙黄酒	2 tbsp yellow wine
3 汤匙生抽	3 tbsp light soy sauce
1.5 汤匙老抽	1.5 tbsp dark soy sauce
15 克冰糖	15g rock sugar
8 个鹌鹑蛋	8 quail eggs
4 个土豆	4 potatoes
盐和胡椒粉	salt and pepper
半盒绢豆腐(大约175克)	half a pack (≈175g) of silken tofu

将猪肉切成1厘米厚的肉片,切掉多余的肥肉。然后将肉片切成细条,切丁。锅中倒入油,用高火加热,再加入猪肉,不停地翻炒,以确保猪肉不粘锅。继续高火翻炒,直到肉中的油充分煸出。

在炒猪肉的时候,可顺便把胡萝卜和洋葱切碎,放入碗中。将生姜切成厚片,并用刀的边缘将皮刮去。当肉中的油充分煸出后,加入切碎的蔬菜和香料(姜、八角茴香、桂皮和月桂叶)。保持高火,时不时地翻炒一下,直到洋葱变得半透明(大约5分钟)。加入黄酒,然后再加入生抽和老抽,搅拌混合。火调小,然后盖上锅盖焖一会儿。用小火煮大约20分钟。煮到一半的时候,查看一下是否煮干了。如果煮干了,加点水或者高汤。

将另一个小锅中的水煮沸。用漏勺小心地将鹌鹑蛋放入水中,确保其完好无损,煮2分钟。盛出鹌鹑蛋,放入冰水中,让鹌鹑蛋迅速冷却。冷却后,剥去鹌鹑蛋的壳。鹌鹑蛋的蛋黄应该还是软的,所以剥壳时要格外小心。将去壳的鹌鹑蛋加入肉酱中,轻轻地混合。关火后,让肉酱慢慢冷却下来。盛出香料(姜、八角茴香、桂皮和月桂叶),然后盖上锅盖。

煮沸另一个大锅中的水,同时将土豆去皮切开。水煮沸后,加入适量的盐和土豆。煮20分钟左右,直到土豆变软。沥干水后,把土豆放回锅里,捣成土豆泥。将豆腐放入另一个碗中,进行搅拌,直到变得细腻。往土豆泥中加入适量的盐和胡椒粉调味,搅拌均匀,然后把所有的食材都加进去。

将烤箱预热。将土豆泥装入一个裱花袋,如果你没有裱花袋的话,一个保鲜袋也可以。将土豆泥挤压到烤盘底部,然后在角上切开一个小口。将肉酱平铺到烤盘上,并确保每个鹌鹑蛋都稳稳地摆放在上面。使用裱花袋或者保鲜袋将土豆泥挤到肉酱上。一旦土豆泥覆盖了烤盘以后,用叉子在上面画几条线,这些之后都会变脆。将烤盘放在靠近热源的架子上,直到土豆泥顶部烤得金黄酥脆。配上武斯特酱(又名英国黑醋)享用,味道极佳。

restaurant
The Table

Modern European, interactive fine dining
现代欧式佳肴

chef
SEAN CLARK

Edinburgh, United Kingdom
爱丁堡，英国

Glasgow Scallops

格拉斯哥扇贝

扇贝
8只带壳的扇贝
2汤匙葡萄籽油

烤花椰菜
12个小花椰菜
20克无盐黄油

花椰菜泥
切好的花椰菜
200克牛奶
50克奶油
1/2茶匙咖喱香料

葡萄干油醋汁
40克黄金葡萄干
40克杏仁
60克油菜籽油
20克麦芽醋

红葡萄酒糖浆
150毫升波特酒

Scallops
8 scallops with shell
2 tbsp grapeseed oil

Roasted Cauliflower Florets
12 small cauliflower florets
20g unsalted butter

Cauliflower Puree
all the cauliflower trimming
200g milk
50g cream
1/2 tsp curry powder

Raisin Vinaigrette
40g golden raisins
40g almonds
60g rapeseed oil
20g malt vinegar

Red Wine Syrup
150ml port wine (Buckfast)

【烤花椰菜】在小煎锅中加入黄油,小火加热。待黄油融化,加入小花椰菜,适当调味后再炒,时不时地轻轻晃动煎锅,炒到菜变软(大约10分钟)。将菜从锅中盛出,放在厨房用纸上备用。

【花椰菜泥】将花椰菜切块,放入平底锅,加牛奶和奶油,再加一点盐调味,煮到软,沥干。将花椰菜放入搅拌机中,加些水(如果需要的话),打成柔滑的菜泥。进行细筛,如有需要,可进行调味。将一半的花椰菜泥和咖喱粉混合。

【葡萄干油醋汁】烤箱预热至120°C。把葡萄干放在一个碗里,用沸水浸没,浸泡10分钟,然后沥干、切碎。用烤箱低温烘烤杏仁大约6分钟,然后切碎。最后把葡萄干、杏仁、油和醋混合。

【红葡萄酒糖浆】将波特酒倒入平底锅中,用中小火煮,收汁,直到它达到像糖浆一样的稠度。

【扇贝】将扇贝和油倒入煎锅中,用中高火加热1分钟。翻另一面后,再加热30秒,从锅中取出。适当调味后进行保温。

【装盘】在每个扇贝壳中盛一勺花椰菜泥,平铺,淋上红葡萄酒糖浆,旁边配上烤好的花椰菜。将两个扇贝放在花椰菜上,然后淋上葡萄干油醋汁,即可享用。

**Chateau d'estoublon
extra virgin olive oil
single variety beruguette**

圣图德斯布隆
特级初榨橄榄油

橄榄油是西方烹饪中经常用到的油，而特级初榨橄榄油则主要用于沙拉和装盘。

**URBANI
black truffle drops
olive oil**

黑松露橄榄油

只用不加盐的黄油来烹饪。

**Elle & Vire
unsalted butter**

爱乐薇淡味黄油块

**Red & white
wine vinegar**

红、白葡萄酒醋

--- Day 5 ---

HEART
用心

| JUAN GOMEZ | KOEN VESSIES | EL WILLY |

今天我学到了这样一个道理：一道用当地新鲜的食材做的简单菜肴，当你用爱和热情来烹饪这道菜肴的时候，总比漫不经心地烹饪复杂的菜肴要好。食物的能量来自于创造者，当你吃的时候，你绝对可以品尝出这细微之处。

无论是做一碗面条，一份简单的沙拉，还是其他美食，都让我们一起用心、用爱去烹饪吧！

chef
JUAN GOMEZ

Mexico City, Mexico

墨西哥城，墨西哥

restaurant
Tomatito
"Sexy Tapas Bar"

Traditional and contemporary Spanish tapas

传统与当代相结合的西班牙佳肴

Balik Salmon

Balik牌三文鱼

85克烟熏三文鱼
5克黑松露

85g smoked salmon
5g black truffle

迷你皮塔面包
320克面粉
190毫升牛奶
5克盐
10克干酵母

Mini Pita Bread
320g flour
190ml milk
5g salt
10g dry yeast

松露酸奶油
500克轻质酸奶油
100克松露油
200克蜂蜜

Truffle Sour Cream
500g light sour cream
100g truffle oil
200g honey

松露蜂蜜
200克蜂蜜
100克松露油

Truffle Honey
200g honey
100g truffle oil

【迷你皮塔面包】烤箱预热至220°C。把所有准备做皮塔的食材搅拌在一起，揉成一个面团。在面团上盖上保鲜膜，让它在一个暖和的地方静置4小时。面团发酵好后，用擀面杖擀平面团，把它切成一个个直径为4厘米的圆形。放入烤箱烘烤3—4分钟，使其发酵并且呈金黄色。然后从烤箱中取出，让其冷却。

【松露酸奶油】将所有的原料放在碗里，用搅拌机进行搅拌，直到混合物变得细腻。

【松露蜂蜜】将两种原料放在碗里搅匀。

【烟熏三文鱼】将三文鱼切成长方形的片状，然后卷起来。

【装盘】用裱花袋装松露酸奶油并将其填入皮塔面包中。将面包放在餐盘上，再将三文鱼卷放在面包上面，用黑松露点缀每一片三文鱼，最后把松露蜂蜜淋在三文鱼上。

chef
KOEN VESSIES

Leiden, Netherlands

莱顿，荷兰

restaurant
El Willy

Fun and casual approach to contemporary Spanish cuisine, sharing style

轻松愉悦的用餐环境
当代西班牙佳肴

Lemon Cream with Berries, Lime Jelly & Basil Sorbet

柠檬奶油配浆果，
青柠果冻配罗勒冰糕

柠檬奶油	**Lemon Cream**	**橙子蛋白酥**	**Orange Meringue**
600 克柠檬	600g lemon	200 克蛋白	200g egg white
150 克白砂糖	150g sugar	300 克白砂糖	300g sugar
140 克蛋黄	140g egg yolk	100 克细砂糖	100g fine sugar
160 克鸡蛋	160g eggs	400 克橘子	400g orange
75 克明胶片	75g gelatin sheet		
150 克黄油	150g butter	**咸饼干碎**	**Salt Crumble**
		250 克白砂糖	250g sugar
青柠果冻	**Lime Jelly**	250 克杏仁粉	250g almond powder
150 毫升水	150ml water	250 克面粉	250g flour
150 克白砂糖	150g sugar	20 克盐	20g salt
75 克青柠汁	75g lime juice	1 千克橘子，榨成汁	1kg oranges, juiced
15 克明胶片	15g gelatin sheet		
罗勒冰糕	**Basil Sorbet**		
625 毫升水	625ml water		
650 克青柠	650g lime		
185 克白砂糖	185g sugar		
50 克葡萄糖	50g glucose		
50 克罗勒	50g basil		

【柠檬奶油】将明胶片放在冰水里,静置5分钟,直到变软。将柠檬挤出汁。将柠檬汁和白砂糖、蛋黄及整个鸡蛋搅拌在一起,然后倒入双层锅加热并且一直搅拌,直到混合物呈现奶油状的质地,取出明胶,沥干,加入混合物中。缓慢地加入块状黄油直到其充分融化。将混合物过筛后放进冰箱里储存。

【青柠果冻】将明胶片放在冰水里,静置5分钟,直到变软。将剩余的所有制作青柠果冻的食材混合在一个锅里,然后加热、煮沸。沸腾后,将锅从炉子上移开。将沥干的明胶加入混合物中,搅拌混合物直至明胶完全融化。将混合物放在一个平的容器里,然后在冰箱里静置2小时,使其凝固。

【罗勒冰糕】将所有的食材混合在一起,然后煮沸。一起搅拌后,过筛。接着放进冰箱冷冻室里,确保每30分钟从冷冻室里拿出来搅拌一次,直到它变成冰糕状。

【橙子蛋白酥】将蛋白放入厨师机中,低速打发1分钟后,逐渐提速,继续打发,直到蛋白的体积变成了原先的两倍。每次逐步加入少许白砂糖,再将蛋白打至发亮、非常结实(硬性发泡)的蛋白霜。将蛋白霜从厨师机中取出,放入已经预热至100°C的烤箱,烤一个半小时。加入用刨刀刨好的橙皮,储存在冷藏室里。

【咸饼干碎】混合所有食材,用你的手掌揉面,直到揉出湿润的面团。将面团放进100°C的烤箱里,烘45分钟直至变得金黄、松脆,再用手碾碎。

【装盘】从冰箱里取出果冻并切成块状。将柠檬奶油挤在餐盘的底部,并放上一些切好的果冻,用蛋白酥进行点缀,再撒一些咸饼干碎和混合的浆果进行装饰,最后加一勺冰激凌,即可享用。

Follow me

Heidi 跟我学

我们有
- 最棒的厨师
- 分步骤讲解的视频和图片

Our
- Best chefs
- Step by Step videos and photo guides

 Heidi 跟我学

 Heidi 中洋生活

 Heidi 中洋生活

今天就加入"Heidi 跟我学"大家庭吧！

chef
EL WILLY

Barcelona, Spain

巴塞罗那，西班牙

restaurant
El Willy

Fun and casual approach to contemporary Spanish cuisine, sharing style

轻松愉悦的用餐环境
当代西班牙佳肴

Scallop Ceviche, Avocado & Crispy Shallots

葱香扇贝配牛油果慕斯

牛油果慕斯

3 个牛油果
8 克红辣椒
3 克香菜
3 克青柠汁
2 克塔巴斯科辣酱
2 克伍斯特辣酱油
1 克盐
15 克葱

扇贝

70 克扇贝
20 克蚝油
15 克葱
1 克芥菜
2 克盐和 2 克黑胡椒粉
5 毫升特级初榨橄榄油
1 克水芹、细叶芹、香菜
1 朵黄瓜花
2 朵秋海棠（花）

Avocado Mousse

3 avocados
8g red chili
3g coriander
3g lime juice
2g Tabasco sauce
2g Worcestershire sauce
1g salt
15g shallot

Scallops

70g scallop
20g oyster sauce
15g shallots
1g mustard greens
2g salt and 2g black pepper
5ml extra virgin olive oil
1g cress, chervil, coriander
1 cucumber flower
2 begonias (flowers)

先把辣椒去籽，切末。然后香菜摘掉叶子，也切末。将葱切末，青柠榨成汁。把牛油果去皮、去核，放进搅拌机里，打成泥状。将牛油果泥倒入一个碗中，并将剩余的食材也倒进去。品尝后进行调味，如果需要的话，可以再加盐或青柠汁。将果泥放进一个裱花袋或者塑料容器里，冷藏。

将葱切丝，炒香。然后将葱丝盛放在厨房用纸上冷却。

将扇贝洗净并切下扇贝肉。在餐盘的中央，铺上牛油果慕斯，然后再放上10块扇贝肉，铺满盘子。将蚝油淋在扇贝肉上，撒上少许香葱和芥菜，用适量的盐和黑胡椒粉调味。然后用可食用花朵进行点缀，最后淋上橄榄油。

Baby Burrata

迷你布拉塔

番茄果酱	Tomato Jelly
2.5千克番茄	2.5kg tomatoes
135克明胶片	135g gelatin sheet
2克盐	2g salt
8克辣椒酱	8g tabasco
7克糖	7g sugar
30克青柠	30g lime

水牛芝士	Mozzarella
3个水牛芝士小球	3 small mozzarella balls
30克番茄果酱（见上方）	30g tomato jelly (see above)
50克红樱桃番茄	50g red cherry tomatoes
25克黄樱桃番茄	25g yellow cherry tomatoes
25克绿樱桃番茄	25g green cherry tomatoes
5克芝麻菜	5g arugula
5克松仁	5g pine nuts
30毫升橄榄油	30ml olive oil
8毫升意大利香醋	8ml balsamic vinegar
3克盐和2克黑胡椒粉	3g salt & 2g black pepper
1克紫苏水芹	1g purple perilla
1克罗勒水芹	1g purple basil
1朵紫罗勒花	1 purple basil flower
3朵秋海棠	3 begonias
1朵黄瓜花	1 cucumber flower
10毫升特级初榨橄榄油	10ml extra virgin olive oil

【番茄果酱】将番茄切碎，放进碗里，用保鲜膜包紧。同时，将明胶片放入冰水中软化，沥干，隔水加热成液态。用适量的盐、辣椒酱、糖和青柠调味。然后将明胶倒入碗中一起搅拌。接下来放入锅中，让番茄果酱静置凝固。凝固后，分装进保鲜袋里，待用。

将樱桃番茄切丁。烘烤松仁。

【装盘】将樱桃番茄放在餐盘底部，涂上番茄果酱。将水牛芝士小球对半切，围成圈放在番茄上，在每个水牛芝士小球上面放一片芝麻菜。然后撒上少许松仁，淋上意大利香醋和橄榄油，再用适量的盐和黑胡椒粉调味。最后用鲜花进行点缀即可享用。

Day 6
UNIQUE STYLE
独特的风格

| CRAIG WILLIS | KASPER ELMHOLDT PEDERSEN | LEK CHAYOO |

今天学会的最重要的一课就是：忠于自己，创造独特的风格。这样，当你的客人看见、品尝你烹饪的食物时，他们可以识别出这是属于你的标记。每位厨师都有自己独特的风格，都会根据他们自己的经验和喜好，创造出能够展现自己灵魂的菜肴。

摆盘也是一个能够彰显你独特风格的非常重要的环节。对于如何摆盘要从长计议，总的来说，有时候放少一点食物比把全部的食物都装在盘子里好。多余的食物可以放在共享盘里，放在桌子的中间，提供给那些还想再吃点的人。

chef
CRAIG WILLIS

Sydney, Australia

悉尼，澳大利亚

restaurant
mr willis

Modern Mediterranean, rustic
It all began with mr willis: a Shanghai dream of Australian chef Craig Willis to create a cosy loft kitchen-dining room where he could prepare a small seasonal menu for his guests like he would in his own home.

现代地中海式、自然且质朴的菜肴
一切都源于韦栗士先生的一个上海梦：创造一个舒适的阁楼厨房餐厅，他可以为客人准备一个季节性菜单，就像在他自己家里一样。

mr willis Roast Chicken with Provençale Pumpkin

韦栗士烤鸡和普罗旺斯烤南瓜

烤好一只鸡，是每一位家庭厨师都应掌握的技能。

烤鸡	Roast Chicken
1整只鸡（约2千克）	1 whole chicken (≈2kg)
20片干的或新鲜的咖喱叶	20 curry leaves
1个柠檬	1 lemon
2支肉桂棒	2 sticks of cinnamon
盐和胡椒粉	salt & pepper
橄榄油	olive oil
白葡萄酒	white wine

焗南瓜	Pumpkin Gratin
1千克南瓜	1kg pumpkin
4汤匙面粉	4 tbsp flour
4汤匙欧芹，切碎	4 tbsp chopped parsley
120毫升橄榄油	120ml olive oil
3瓣大蒜，拍碎	3 garlic cloves smashed
盐和胡椒粉	salt & pepper

【烤鸡】取一只新鲜的鸡，去掉鸡头、鸡脖和鸡脚，洗净后在鸡的表皮和鸡的内部充分均匀地抹上盐。取四分之一个柠檬，挤出汁淋在鸡上。撒上少量的咖喱叶，放入切成大块的肉桂棒。接着把鸡放在一个餐盘中或者放进一个塑料保鲜袋，然后放进冰箱冷藏12小时或者过夜。

把鸡用厨房细绳绑起来，拉紧成一个球形。撒上适量盐和胡椒粉，用橄榄油均匀地涂抹。然后放进烤盘，也可以把柠檬皮放进去。将鸡放入烤箱，设置200°C烘烤。烤15分钟后，把鸡取出翻面，并再次放入烤箱继续烘烤15分钟。实际的烘烤时间取决于烤箱的功率。可用一把刀插入鸡腿和鸡胸肉的连接处，检查鸡肉是否已经烤熟。烤熟后让鸡在烤盘中静置10—20分钟。

【焗南瓜】南瓜洗净后，去皮，用一个勺子刮去瓤。把南瓜切成稍大的块状，把其余的食材混合在一起，放进烤箱专用盘——陶瓷焗饭盘，就像烤千层面时用到的焗饭盘一样。将烤盘放入预热至180°C的烤箱，烤至南瓜内软外脆。

【汤汁和装盘】把烤好的鸡从烤盘里拿出来。先切下鸡翅，再切鸡腿，最后切鸡胸肉。把烤盘里的汁水倒入一个小平底锅里，然后放在炉子上加热。倒入一杯约50毫升的白葡萄酒。刮一下烤盘里剩余的食材，然后将它们倒入这个平底锅里，加入适量盐和胡椒粉调味。最后把调好的汤汁搭配烤鸡和南瓜一起享用。

Braised Beef Cheeks with Coco & Polenta

可可炖牛肉配意大利玉米饼

牛脸颊肉

4 汤匙橄榄油

4 块牛肉（360 克）

1 个中等大小的胡萝卜，切块

1 个中等大小的洋葱，切片

1/2 根西芹，切成 2 厘米块状

1/2 杯不加糖的可可粉

2 杯红葡萄酒或黑啤

1 罐番茄，切碎

现磨黑胡椒粉和盐

欧芹叶子，去除茎，切碎

玉米粥饼

30 克黄油

30 克帕尔玛干酪

1.2 升冷水

2 茶匙盐

500 克玉米面粉

白胡椒粉

Beef Cheeks

4 tbsp olive oil

4 beef cheeks (360g)

1 medium carrot, diced

1 medium onion, cut into pieces

1/2 celery stick, cut into 2cm pieces

1/2 cup unsweetened coco powder

2 cups red wine or dark beer

1 can chopped tomatoes

ground black pepper & salt

flat leaf parsley, remove stalks and roughly chop

Polenta

30g butter

30g Parmesan cheese

1.2l cold water

2 tsp salt

500g corn flour

white pepper

【牛脸颊肉】把烤箱预热至180°C。用厨房专用纸把牛脸颊肉擦干，涂抹适量的盐和胡椒粉。在砂锅中加入两汤匙橄榄油后，调至高火加热，油热后倒入牛脸颊肉，煎至变色。然后翻一面，继续煎。把牛脸颊肉从锅里取出来，然后放在一边静置。将锅里剩余的油脂以及橄榄油倒尽。

再向锅里加一些橄榄油，把火调至中火。翻炒洋葱、胡萝卜和西芹，炒至变软。松软后，立即加入可可粉和酒，直至汤汁剩下一半（让酒精挥发掉，不然会有苦涩味）。将牛脸颊肉回锅，再加一些番茄，加入适量的盐和黑胡椒粉。等汤汁烧开后，用锡纸封口，盖上锅盖。放进烤箱，调至180°C，烘烤3小时，直到牛脸颊肉变得酥软。

【玉米粥】在小锅中烧水，加一点盐。水煮沸后，把锅从炉子上移开，慢慢地倒入玉米面粉，同时不断搅拌直至均匀。把锅放回炉子上，调至小火继续加热，并不停地搅拌约20分钟。再加入适量的帕尔玛干酪、盐和白胡椒粉，继续搅拌。

【玉米粥饼】将黄油涂抹在一个正方形的托盘上。再将调味后的玉米粥均匀地倒入托盘，包上保鲜膜，放进冰箱。待玉米粥冷却后，从冰箱里取出。把玉米粥切成三角形。在小煎锅中倒入橄榄油后加热。油热后，把成形的玉米粥三角饼两面都煎一下，直至表面酥脆。从锅中取出待用。

【装盘】将牛脸颊肉和汤汁盛入一个餐盘，配上三块玉米粥三角饼，撒上少量欧芹即可享用。

restaurant
Pelikan

Casual Nordic Dining
Nordic cuisine is renowned worldwide for its innovation and finesse using a wide range of culinary techniques.

北欧休闲餐厅
北欧美食以创新和巧妙的烹饪技术享誉全球。

chef
KASPER ELMHOLDT PEDERSEN

Esbjerg, Denmark
埃斯比约，丹麦

这是最美味的沙拉! ♥

Thinly Sliced & Rolled Zucchini with Spiced Dill Cream

薄片西葫芦卷配莳萝奶油酱

2 个西葫芦	2 zucchinis
2 捆新鲜莳萝	2 bundles of fresh dill
1/2 捆葱	1/2 bundle chives
2 根腌制的甜菜根（切薄片）	2 pickled beetroots (slice thinly)
1 个柠檬	1 lemon
橄榄油	olive oil

酱汁	Dressing
200 克奶油芝士	200g cream cheese
100 克奶油	100g cream
50 毫升醋	50ml vinegar
1 茶匙辣椒粉	1 tsp cayenne pepper powder
20 克切碎的生姜	20g ginger chopped
2 茶匙香菜籽粉	2 tsp coriander seed powder
2 茶匙五香粉	2 tsp five spice powder
50 克切碎的莳萝	50g chopped dill

【莳萝奶油酱】把所有的食材放入碗中，搅拌均匀。品尝一下味道，再放入盐和柠檬汁调味。

【装盘】在餐盘的中间放一点酱，平铺成0.5厘米厚度的大型圆状。把西葫芦切成薄长条片。把每一条薄片西葫芦卷成宽松状，放在莳萝酱上面，把腌制的甜菜根片放在西葫芦卷上。然后在沙拉盘中放入切碎的莳萝和葱，再撒上少许柠檬皮，淋上橄榄油，便可享用。

Strawberries & Rhubarb

草莓大黄酥

草莓果酱

1千克草莓

1.5千克糖

1个大的玻璃罐

大黄

4根新鲜大黄

100毫升水

饼干块

4个鸡蛋

70克细砂糖

70克室温下的软黄油

115克面粉

1克盐

8克泡打粉

抹面

你最喜欢的香草冰激凌

酸奶油

Preserved Strawberries

1kg strawberries

1.5kg sugar

1 large glass jar

Rhubarb

4 sticks of fresh rhubarb

100ml water

Crumble

4 eggs

70g fine sugar

70g soft butter at room temperature

115g flour

1g salt

8g baking powder

Plating

your favorite vanilla ice cream

sour cream

【草莓果酱】在玻璃罐里装满草莓，让草莓100%完全覆盖上细砂糖。12小时之后，查看一下，再加点细砂糖以确保草莓被完全覆盖。将装满草莓的玻璃罐在室温下放置4天（不要放进冰箱）。4天之后，果酱就做好了，保留备用。

【大黄】把大黄洗净后削皮，切成5厘米厚的块状，然后放进一大锅煮沸的水里煮软。将煮好的大黄从水里捞出，沥干，然后放在一旁。

【饼干块】把烤箱预热至160°C。把所有的食材混合在一起，放在烤盘上，烤30—40分钟。将烤好的饼干块从烤箱里取出来，放在一旁。

【抹面】把酸奶油涂抹在餐盘上，再放上草莓果酱、大黄以及饼干块，最后淋上一大勺香草味冰激凌。

Tom Yum Seafood Soup

冬阴功海鲜汤

★ 我最爱用这款冬阴功汤作为火锅的汤底。♡

海鲜汤底
- 5 个虾头
- 60 克冬阴功酱
- 10 克虾干
- 100 克洋葱，切片
- 50 克香茅，切碎
- 10 克南姜，切片
- 50 克胡萝卜，切片
- 100 克番茄，切片
- 3 片青柠叶
- 4 个红辣椒，分别切成两段
- 1.7 升水
- 30 克鱼露
- 80 克青柠檬汁
- 8 克盐
- 20 克糖
- 鸡精
- 50 克番茄沙司
- 素油

Seafood Broth
- 5 shrimp heads
- 60g tom yum soup paste
- 10g dry shrimp
- 100g onions sliced
- 50g lemongrass chopped
- 10g galangal sliced
- 50g carrot sliced
- 100g tomatoes sliced
- 3 lime leaves
- 4 red chilies chopped
- 1.7l water
- 30g fish sauce
- 80g lime juice
- 8g salt
- 20g sugar
- chicken powder
- 50g tomato paste
- vegetable oil

做汤食材
- 4 只大明虾
- 12 个蛤蜊
- 8 个鱿鱼
- 20 个草菇
- 8 片番茄，切片
- 4 个红辣椒
- 4 片青柠叶
- 8 克香菜
- 8 克炼乳
- 2 个泰国青柠檬

Filler for Soup
- 4 king prawns
- 12 clams
- 8 squid
- 20 straw mushrooms
- 8 tomato slices
- 4 red chilies
- 4 lime leaves
- 8g coriander
- 8g evaporated milk
- 2 Thai limes

【汤底】在大锅里加热素油。油加热后，加入虾头炒香，再加入冬阴功酱和虾干，炒2分钟。加入切片的洋葱、香茅、南姜、胡萝卜和番茄，迅速搅拌食材，然后加水，煮沸后，调至小火炖30分钟。在锅里加入鱼露、青柠檬汁、盐、糖、鸡精、番茄沙司和香菜，再煮沸，然后关火，接着把所有的食材从汤里过滤出来。

【海鲜】把过滤好的汤再回锅煮沸，加入准备好的海鲜、草菇、番茄、红辣椒、青柠叶和香菜。当海鲜刚煮熟的时候，加入炼乳和泰国青柠檬。把汤盛入汤碗中，撒上少许香菜便可享用。

**KOTÁNYI
Ground Chili Pepper**
可达怡辣椒粉

Salt
盐

Star anise
八角茴香

Fresh red chili
鲜红辣椒

**KOTÁNYI
Multicolored
Pepper**
可达怡多彩胡椒粒

Dry red chili
干红辣椒

Bay leaf
月桂叶

Thyme
百里香

如果你想烹饪
国际美食,
家里存储一些晒干的
香草和食材
是很有必要的。

Turmeric
鲜姜黄

Rosemary
迷迭香

KOTÁNYI
Oregano
Chopped
可达怡牛至叶碎

Basil
罗勒叶

Clove
丁香

Sage
鼠尾草

Day 7

EXPERT
成为专家

| WILLIAM WANG | WEI CHEN | WILLMER COLMENARES |

你不需要把任何事都做得尽善尽美。事实上,你也不可能。但是你可以选择一些自己擅长的领域,争取在这些领域做到最好。

今天这些厨师在他们的领域中是首屈一指的。他们把时间花在磨炼自己的技能上,学习自己需要知道的一切。与这样的人一起工作,可深受启发。

chef
WILLIAM WANG

Shanghai, China

上海，中国

brand
ROUGIÉ

Founded in 1875 in France, Rougié has become a great choice as the Foie Gras supplier toward hotels, restaurants and catering chefs in the world.

露杰，1875年成立于法国，作为鹅肝酱的供应商，它已经成为世界各地酒店、餐厅和餐饮大厨的绝佳选择。

★ 关注微信公众号"Heidi跟我学"，
查看如何烹调你的鸭胸脯，使它不油腻。

Roasted French Style Duck Breast with Herb Crust & Mashed Potato

香烤法式鸭胸配
松露土豆泥

350克法式鸭胸	350g French style duck breast
50克面包屑	50g bread crumbs
100克土豆	100g potatoes
2克百里香	2g thyme
2克迷迭香	2g rosemary
5克欧芹	5g parsley
苗菜	microgreens
5克芥末酱	5g mustard
10克黄油	10g butter
50毫升牛奶	50ml milk
少许松露油	a little truffle oil
盐和黑胡椒粉	salt & black pepper

把面包屑、百里香、迷迭香和欧芹放在搅拌机里搅碎，作为香料备用。

处理鸭胸。先在鸭胸表皮上划几刀，然后放在煎锅里，小火烹调，直至表皮呈现棕色，然后翻一面煎一会儿，再翻回来，大约煎11分钟后，将鸭胸拿出来，在无皮的那一面抹上芥末酱，再用香料包裹住鸭胸肉，放入烤箱190°C烤3分钟。最后切片装盘。

在餐盘边上倒一些红葡萄酒汁，放点土豆泥，用少许苗菜装饰摆盘，然后在肉上撒上适量的盐和黑胡椒粉，即可享用。

chef
WEI CHEN

Shanghai,
China

上海，中国

company
SINODIS

Good food, good living.

西诺迪斯
中国领先的食品经销商

chef
WILLMER COLMENARES

Caracas, Venezuela
加拉加斯，委内瑞拉

restaurant
CHARdining
Hotel Indigo Shanghai On The Bund

Premium Steakhouse
高级牛排馆

CHAR Grilled Tomahawk

炭烤战斧

1.3千克战斧牛肉	1.3kg Tomahawk
100克无盐黄油	100g unsalted butter
2根迷迭香	2 sprigs of rosemary
5根百里香	5 sprigs of thyme
2个大蒜	2 garlic cloves

白大蒜酱 / White Garlic Puree

1千克大蒜，去皮	1kg garlic peeled
200克黄油块	200g butter block
10克盐	10g salt

炒蘑菇 / Sautéed Mushrooms

300克时令蘑菇	300g seasonal mushrooms
10克大蒜，切碎	10g garlic chopped
5克欧芹叶子，切碎	5g parsley leaves chopped
25克黄油	25g butter
松露油	truffle oil

樱桃番茄 / Cherry Tomatoes

300克樱桃番茄，对半切	300g cherry tomatoes cut in halves
20克大蒜，切片	20g garlic sliced
10根百里香	10 sprigs of thyme
盐、胡椒粉、糖	salt, pepper, sugar
橄榄油	olive oil

在你开始烹饪之前，先从冰箱里取出战斧牛肉解冻，静置30分钟。烤箱预热至80°C。

【樱桃番茄】切掉番茄的两头，然后对半切。将番茄放在一个托盘上，切开的一面朝上方，撒上适量的盐、胡椒粉、糖和切碎的百里香，淋上橄榄油。放进烤箱烤45分钟，待番茄变软后，从烤箱中取出，放在一边让其冷却，待用。

【烤牛肉】烤箱调至180°C。尽可能加热烤架（至少200°C）。用盐腌制牛肉，将橄榄油均匀地刷在牛肉上以确保烤的时候不会黏住烤架。一边烤大约2分钟后，在同一侧转90度，再烤2分钟，然后再转动，重复这个过程，直到牛肉的两面都有像钻石形状的烧烤印记。将牛肉从烤架上取下来并放进烤箱。烤大约35分钟，每8分钟左右将牛肉翻面，以确保两面烤得一致。将牛肉从烤箱中取出，在装盘前静置15—20分钟，这样会使牛肉颜色很好看，还有肉汁。

【白大蒜酱】在炖锅中加水，放入去皮的大蒜一起煮。煮沸后，取出大蒜将水沥干，再往锅里加冷水。水沸后，将之前的大蒜放入，继续煮30分钟。关火，将水沥干。将大蒜放入搅拌机，低速搅拌均匀。分次加入一块块的黄油，至混合物发白、呈现柔滑状态即可。加盐适当调味，然后放在一旁备用。

【炒蘑菇】清洗蘑菇（切记不要直接用水洗，用湿布，如有需要可用刷子），将蘑菇切成大小均匀的形状。在煎锅中加入一点橄榄油并加热，慢慢地加入蘑菇，不停地翻炒，直到蘑菇呈现炒过的颜色。再加一小块黄油、切碎的大蒜和酱汁，将蘑菇炒熟。最后加入切碎的欧芹叶子，关火。

【装盘】如果战斧牛肉已经凉了，放回烤箱再烤5分钟回热（不是继续烤）。将战斧牛肉均匀地切好放在餐盘上，淋上几滴蒜酱，在边上搭配蘑菇即可一起享用。

每次我看到这张图的时候，我就感觉饿了。

Smoked Chocolate Temptation

熏巧克力诱惑

巧克力翻糖	Chocolate Fondant
115 克 66% 黑巧克力	115g 66% dark chocolate
105 克黄油	105g butter
30 克蛋黄	30g egg yolk
90 克蛋白	90g egg white
40 克面粉	40g flour
50 克白糖	50g sugar

辣黑巧克力酱	Spicy Dark Chocolate Ganache
200 克黑巧克力	200g dark chocolate
200 克奶油	200g cream
40 克咖啡利口酒	40g coffee liquor
30 克陈年朗姆酒	30g dark aged rum
1 克辣椒粉	1g cayenne pepper
2 克速溶咖啡	2g instant coffee

你最喜爱的香草冰淇淋　your favourite vanilla ice cream
100克去壳的开心果　100g peeled pistachios

榛子果仁糖　**Hazelnut Praline**
65克糖粉　65g icing sugar
75克黄油　75g butter
20克玉米糖浆　20g corn syrup
40克面包粉　40g bread flour
40克榛子，轻轻粉碎　40g hazelnuts lightly crush

海盐焦糖酱　**Salted Caramel Sauce**
100克糖　100g sugar
100克奶油　100g cream
50克黄油　50g butter
5克盐　5g salt

【巧克力翻糖】在模具里涂上黄油。将巧克力和黄油隔水熔化，然后放置在一旁。在另一个碗中打发蛋白，少量多次地拌入白糖，直到打成柔软的蛋白霜，再加蛋黄和面粉，慢慢地用塑料刮刀搅拌，不要过度搅拌，然后放进冰箱里备用。

【辣黑巧克力酱】隔水熔化巧克力。熔化后，从炉上移开，慢慢地加入酒、香料和奶油，直至柔滑。冷却后放入冰箱，待装盘用。

【海盐焦糖酱】将锅中的糖和水用中火加热，直到变成焦糖色。慢慢将奶油打入热焦糖混合，不断搅拌，以免过稀。加入黄油和盐。继续用小火煮约15分钟，直到混合物变成略厚的焦糖酱。

【榛子果仁糖】烤箱预热至180°C。把除了坚果之外的所有食材都混合在一起，直到变成光滑的糊状物，平铺在硅胶垫上，将坚果撒在混合物上，放进180°C烤箱，烘烤10分钟，然后从烤箱中取出并让其冷却下来。

【开心果粉】将去壳的开心果放入搅拌机中粉碎，打成绿色粉末。

【装盘】从冰箱中取出巧克力翻糖，在预热180°C的烤箱中烘烤9分钟（烘烤时间取决于烤盘的大小）。取出，翻动一下烤盘，以便让翻糖脱模。然后装盘，把剩余的食材也放在盘子里，与冰淇淋一起享用。

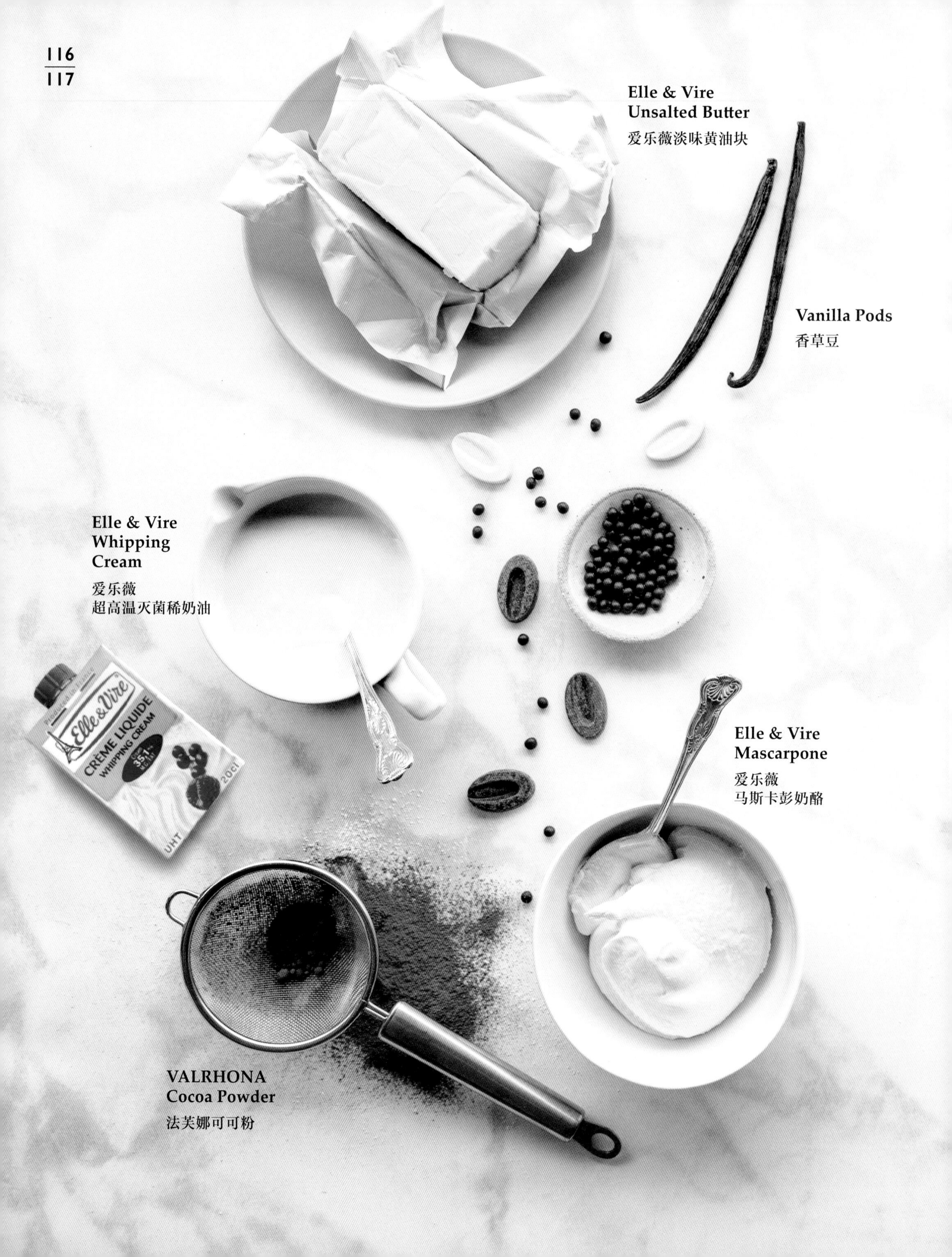

Elle & Vire
Unsalted Butter
爱乐薇淡味黄油块
Vanilla Pods
香草豆
Elle & Vire
Whipping
Cream
爱乐薇
超高温灭菌稀奶油
Elle & Vire
CREME LIQUIDE
WHIPPING CREAM
20cl
UHT
Elle & Vire
Mascarpone
爱乐薇
马斯卡彭奶酪
VALRHONA
Cocoa Powder
法芙娜可可粉

Day 8

HAVE FUN
玩得开心

| EDWARD MAIR | RICK BARTAM | SCOTT MELVIN |

这是我为期八天的烹饪之旅的最后一天以及最后一堂课。烹饪应该是种乐趣，你认真对待的同时，也可以乐享其中。当烹饪变得不再有趣时，那就是到了应该外出就餐，让别人做菜给你吃的时候啦！

hotel
Four Seasons Hotel Shanghai

Cuisines and ingredients from around the world

各国料理和食材

chef
EDWARD MAIR

Glasgow, Scotland, United Kingdom

格拉斯哥，英国

Scottish Shortbread with Lemon Curd Dip

苏格兰酥饼配柠檬酱

酥饼	Shortbread
300 克面粉	300g plain flour
150 克玉米粉	150g corn flour
300 克黄油	300g butter
150 克糖	150g sugar
2 个橙子, 刨橙皮	2 oranges, zested

柠檬酱	Lemon Curd
450 毫升柠檬汁	450ml lemon juice
350 克糖	350g sugar
400 克鸡蛋	400g eggs
20 克柠檬皮	20g lemon zest
300 克黄油(室温)	300g butter (room temperature)

【酥饼】烤箱预热至150°C。在开始制作前,先将黄油从冰箱中取出,在室温下放置30分钟。将软化的黄油及糖放在一个容器中,用中速搅拌,直到黄油变成浅黄色,所有糖都充分溶化。然后再慢慢地倒入面粉和玉米粉,继续搅拌,直到充分混合。最后加入橙皮,搅拌1分钟。
在烤盘中将面团擀开,用叉子标出印记。把面团放入预热的烤箱中烘烤大约20分钟,然后取出,在冷却前,撒上糖,切成手指长短的长条。

【柠檬酱】在锅中加入柠檬汁、糖和柠檬皮,用中高火煮,煮沸后从炉子上移开。在碗里打入一个鸡蛋,慢慢倒入沸腾的柠檬汁,并非常快速地搅拌以确保鸡蛋不会变熟。再将混合物倒回锅中,用小火一边加热一边搅拌。当混合物变得浓稠时关火。等柠檬酱冷却到50°C后再加入黄油,随后倒入一个碗或玻璃罐中。将酥饼放入盘中,边上放上柠檬酱,即可蘸柠檬酱食用。

chef
RICK BARTAM

Sunderland, England, United Kingdom

桑德兰，英国

hotel
InterContinental Shanghai Hongqiao NECC

International cuisine

各国料理

Tips: 可以搭配一片面包或一碗米饭食用。

North Sea Plaice, Scottish Lobster with English Peas & Butter Sauce

北海鲽鱼拼苏格兰龙虾

2 整块鲽鱼，切片（800—1000 克，保留鱼骨熬鱼高汤）	2 whole plaice fish, filleted (800–1000g, reserve bones for fish stock)
2 条苏格兰龙虾尾，洗净	2 Scottish lobster tails, cleaned
3 个小土豆，煮熟、去皮、对半切	3 baby potatoes, boiled and cooked, skin removed, halved
20 毫升特级初榨橄榄油	20ml extra virgin olive oil
1/2 杯冷冻豌豆	1/2 cup frozen peas
豌豆芽	pea sprouts
1 个橙子的汁	juice of one orange
辣椒粉	cayenne pepper
100 毫升鱼高汤	100ml fish stock
50 克无盐黄油	50g unsalted butter
盐	salt

【准备】烤箱预热至180℃。在鲽鱼上撒上少量的盐和辣椒粉进行腌制。将鱼排折两次，形成三层。

【制作】将橄榄油倒入平底锅中，用中火加热，放入龙虾尾翻炒，直到每一面都轻微变色。加入鲽鱼排，面朝下，烹饪1分钟左右，将鱼排翻面。在平底锅中加入橙汁和鱼高汤。加入土豆，将平底锅放入烤箱，烤5—6分钟。把平底锅从烤箱中取出。将鲽鱼排、龙虾尾和土豆平分，放在两个碗中。将剩余的鱼高汤放在炉子上加热，加入黄油直到融化。用滤网将酱汁过滤，重新加热，加入豌豆继续加热1—2分钟。

【装盘】将酱汁和豌豆用勺子淋在鲽鱼和龙虾上，撒上少许豌豆芽，即可享用。

Best of British Event
《芙伦精选》录制活动现场

chef
SCOTT MELVIN
Dundee, Scotland, United Kingdom

邓迪，英国

restaurant
The Commune Social

An eclectic menu of modern tapas, desserts and a well-curated wine, beer and cocktails list

多元化的菜单包含了现代西班牙小吃、甜品和精心设计的酒单

Pork Tenderloin, Roasted Endive, Braised Red Cabbage & Pickled Dates

里脊肉配菊苣甜酸枣

1 块猪里脊	1 pork tenderloin
甜酸枣	**Sweet and Sour Dates**
200 克枣	200g dates
100 毫升米酒醋	100ml rice wine vinegar
100 毫升水	100ml water
100 克糖	100g sugar
一撮盐	1 pinch of salt

红椰菜泥	Red Cabbage Puree
1颗中等大小的红椰菜	1 medium sized red cabbage
4个苹果	4 apples
200克葡萄干	200g sultanas
1瓶红酒	1 bottle of red wine
1瓶波特酒	1 bottle of port wine
1茶匙肉桂粉	1 tsp cinnamon
1茶匙五香粉	1 tsp five spice powder
250克黄油	250g butter
赤霞珠醋或红酒醋	Cabernet Sauvignon vinegar or red wine vinegar
蜂蜜	honey

烤菊苣	Roasted Endive
20颗菊苣	20 endives
10支百里香	10 sprigs of thyme
3—4瓣大蒜切成薄片	3–4 garlic cloves thinly sliced
100克橄榄油	100g olive oil
一撮盐	1 pinch of salt
2—3片月桂叶	2–3 bay leaves

【甜酸枣】需提前一天制作。将米酒醋、水、糖和盐放在一个小的平底锅里煮沸。当糖溶化后，把锅移开，将枣子放入热的液体里。然后将锅中的混合物倒入一个玻璃瓶中。冷却后，将其放入冰箱。

【猪里脊】将猪里脊放入真空袋抽去空气，然后放入65°C的水中，煮30分钟。

【红椰菜泥】烤箱预热至160°C。将红椰菜和苹果切成薄片。在一个大的平底锅内放入红酒、波特酒和香料一起煮沸。再加入红椰菜、苹果、葡萄干和黄油，加入适量的盐和胡椒粉调味。用锡纸盖住，放入烤箱烘烤两小时。烤熟后，将平底锅从烤箱中取出放在炉子上，将多余的汁水倒出。最后用醋和蜂蜜调味，品尝下味道，混合后过筛。

【烤菊苣】将菊苣切成两半，与其他食材充分混合，在预热至160°C的烤箱中烘烤10—12分钟，时不时地进行翻面。
在平底锅中倒入橄榄油并加热，将猪肉从密封袋中取出，放在锅中，煎至变色。将猪肉从锅中取出，切成块。

【装盘】将红椰菜泥放在盘里，加上三块猪肉，再放上烤好的菊苣和腌好的枣。

掌握最新资讯

Keep up to date

每周更新的食谱

With new recipes weekly on Wechat, Weibo

您可以与我随时联系
我很期待您烹饪的美食
可在您编辑的图文中 @ 我

You can connect with me anytime. I'd love to see what you cook — tag me in your photos.

- Heidi 跟我学
- Heidi 中洋生活
- Heidi 中洋生活

今天就加入
"Heidi 跟我学"
大家庭吧!

SPECIAL THANKS
特别感谢

撰写这本书的想法来自我和朋友一起喝咖啡的时候。我很庆幸当初努力去实现了这个想法，它使我和许久未见面的厨师和朋友又再次联系到了一起。我决定享受这个写书的过程，所以我只和那些我认同的人和公司进行合作。

我非常感谢来自以下公司的支持：
【SINODIS】西诺迪斯：你们太棒了！我总算知道为什么厨师们对你们品牌的产品特别偏爱了，你们出色的食材对一道菜有着举足轻重的影响！

【KitchenAid】凯膳怡：本书食谱中涉及的厨师机和料理机都是凯膳怡的产品。我在自己的家庭厨房中也使用它们，它们让生活变得如此简单。这两台机器有我所需要的一切。

【World Kitchen】康宁：在最后关头伸出了援手，给我提供了康宁餐具、晶彩透明锅、瑞仕钻锅、康宁瑰宝锅及刀具等产品，基本上满足了我和厨师们拍摄做菜的所有需求。

我还想特别感谢英国驻上海总领事馆，英国国际贸易部邀请我和我的团队参加《英伦精选》现场活动的录制，使我们能拍摄本书中涉及的来自英国的大厨们！

The idea of writing this book came to me over a coffee with a friend. I'm so happy I pursued it; it has reconnected me with chefs and friends I hadn't seen for awhile. I decided that I wanted to enjoy the process so I have only worked with people and companies that I really like and believe in.

Thank you for the support I received from these companies:
The team at SINODIS. You are amazing. I can see why chefs prefer the products and ingredients from your brands; they really are outstanding and make a huge difference to the end result of a dish.

KitchenAid: The recipes in this book that use blenders and mixers are all created for and with KitchenAid products. I use them in my own home kitchen and they make life so much easier. I have everything I need in two machines.

World Kitchen: Came to my aid at the very last minute and supplied me with Corelle plates, Swiss Diamond and Revere pans, Visions and Corningware pots, knives, cutting boards, electric cooker... Basically everything I needed to film the chefs making the dishes.

I'd also like to give a special thank you to the UK Department of International Trade, Shanghai Consulate, for inviting me and my crew to the *Best of British* Event to film and photograph the British chefs in this book.

MEASUREMENTS
计量单位对照

烹调和食谱均用干量和液量计量单位。所用计量单位或是"杯"和"杯"的一部分（fractions of a cup），或是"汤匙"和"茶匙"。计量用的"杯"和"匙"可在超级市场、百货商店和折扣商店买到。一般的饮用杯和食用匙不能替代这类计量用具。烹调和食谱经常使用缩写。现将常用容量单位及其缩写列出。

1 tbsp = 1 tablespoon（汤匙）= 3 teaspoons（茶匙）
1 tsp = 1 teaspoon（茶匙）= 5 milliliters（毫升）
1 c = 1 cup（杯）= 16 tablespoons（汤匙）
1 lb = 1 pound（磅）= 16 ounces（盎司）= **16 oz**

OVEN TEMPERATURE
烤箱温度对照

摄氏度	100°C	120°C	140°C	150°C	160°C	180°C	190°C	200°C	220°C			
华氏度	200°F	250°F	275°F	300°F	325°F	350°F	375°F	400°F	425°F			
摄氏度	110°C	130°C	140°C	150°C	170°C	180°C	190°C	200°C	220°C	230°C	240°C	250°C
燃气灶刻度	1/4	1/2	1	2	3	4	5	6	7	8	9	10

PHOTOGRAPHY
图片拍摄

website: www.shannondeutrom.com

Shannon Deutrom

她在一个多元文化的家庭中成长，从小接触国际文化和美食，现在她将这些国际元素融入到她独一无二的品牌中呈现。

她从业20多年，作为一名平面设计师，她曾为澳大利亚和伦敦名列前茅的杂志和广告公司工作，专攻生活美学类出版物的设计。后来，她受到启发，将职业重新定位为一名造型设计师，专注于食物的造型和拍摄。

Growing up in a multi-cultural family, from a young age, Shannon was exposed to international culture and cuisine and now incorporates this into her unique brand of styling.

Shannon's professional career began more than 20 years ago as a graphic designer, working for leading magazines and ad agencies in both Australia and London. Working on both food and lifestyle publications. Shannon has been inspired to re-focus her career as a stylist, specialising in food styling and photography.

图书在版编目（CIP）数据

跟我学做五星大厨／（澳）海蒂（Heidi）著；陆韵菲译. — 杭州：浙江教育出版社，2018.5
ISBN 978-7-5536-7144-4

I. ①跟⋯ Ⅱ. ①海⋯ ②陆⋯ Ⅲ. ①菜谱 Ⅳ. ①TS972.1

中国版本图书馆CIP数据核字(2018)第043843号

跟我学做五星大厨
GEN WO XUE ZUO WUXING DACHU

[澳]海蒂（Heidi） 著
陆韵菲 译

责任编辑：张小飞		美术编辑：韩 波	
文字编辑：潘俊丽		责任校对：陈云霞	
责任印务：吴梦菁			

出版发行：浙江教育出版社
（杭州市天目山路40号 邮编：310013）
图文制作：七月合作社　悦阅
印刷装订：杭州富春印务有限公司
开　　本：889mm×1194mm　1/16
印　　张：9.5　　　　　　　字　　数：190 000
版　　次：2018年5月第1版　　印　　次：2018年5月第1次印刷
标准书号：ISBN 978-7-5536-7144-4
定　　价：88.00元

联系电话：0571-85170300-80928
网　　址：www.zjeph.com
版权所有·侵权必究

CONTENTS
目录

01
FRUITS & VEGETABLES
果蔬类 2

02
SEAFOOD & MEAT
海鲜、肉类 14

03
DRY FRUITS & NUTS
干果、坚果类 18

04
SEASONINGS & OTHERS
调味类及其他 22

05
CREDITS RESTAURANT
餐厅名录 28

06
DIRECTORY OF PRODUCTS
产品名录 32

01

FRUITS & VEGETABLES
果蔬类

番茄 *Tomato*	橙子 *Orange*
菠萝 *Pineapple*	苹果 *Apple*
西瓜 *Watermelon*	柠檬 *Lemon*
香蕉 *Banana*	樱桃 *Cherry*
芒果 *Mango*	桃子 *Peach*

扫一扫，听朗读

枣子
Date

黑莓
Blackberry

椰子
Coconut

青柠
Lime

草莓
Strawberry

芦笋
Asparagus

树莓
Raspberry

土豆
Potato

蓝莓
Blueberry

黄瓜
Cucumber

Bok choy
奶白菜

羽衣甘蓝是我最喜欢的绿色蔬菜之一。

Kale
羽衣甘蓝

Iceburg lettuce
包心菜

Vine tomato
番茄

Soybean sprout
黄豆芽

洋葱 *Onion*	菠菜 *Spinach*
芹菜 *Celery*	竹荪 *Bamboo fungus*
芹菜杆 *Celery sticks*	黄椒 *Yellow chili*
地瓜 *Sweet potato*	青椒 *Green chili*
橄榄 *Olive*	甜玉米 *Sweet corn*

扫一扫，听朗读

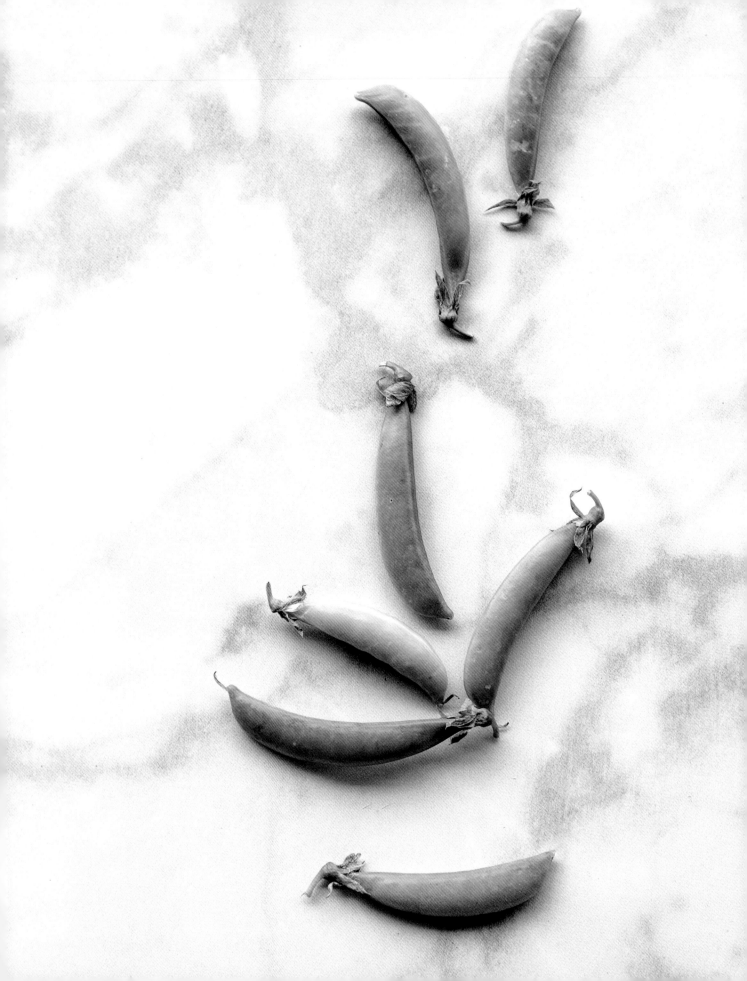

| 青豆 | 南瓜 |
| *Green pea* | *Pumpkin* |

豆芽菜 西葫芦
Beansprout *Zucchini*

甜菜根 卷心菜
Beetroot *Cabbage*

蘑菇 胡萝卜
Mushroom *Carrot*

茄子 韭菜
Eggplant *Leek*

扫一扫,听朗读

02

SEAFOOD & MEAT
海鲜、肉类

牛肉 *Beef*	蟹黄 *Crab roe*
猪肉 *Pork*	扇贝 *Scallops*
羔羊肉 *Lamb*	三文鱼 *Salmon*
鸡肉 *Chicken*	鲈鱼 *Sea bass*
蟹肉 *Crab meat*	鲽鱼 *Plaice fish*

虾
Shrimp

大明虾
King prawn

龙虾
Lobster

蛤蜊
Clam

03

DRY FRUITS & NUTS
干果、坚果类

腰果
Cashew nut

开心果
Pistachio

花生
Peanut

葡萄干
Raisin

无花果
Fig

杏仁
Almond

核桃
Walnut

扫一扫，听朗读

04

SEASONINGS & OTHERS
调味类及其他

| 胡椒粉 | 月桂叶 |
| Pepper | Bay leaf dry |

肉桂
Cinnamon

香菜叶
Cilantro leaf

黄油
Butter

香菜籽
Coriander seed

面粉
Flour

迷迭香
Rosemary

鼠尾草叶
Sage leaf

罗勒
Basil

莳萝
Dill

生抽
Light soy sauce

欧芹
Parsley

老抽
Dark soy sauce

醋
Vinegar

芥末
Mustard

苹果醋
Cider vinegar

酱油
Soy

麦芽醋
Malt vinegar

酱
Soy sauce

扫一扫，听朗读

05

CREDITS RESTAURANT
餐厅名录

chef	restaurant		
Tony Su	**Ming Court** ｜ 明阁		
p038	B1, Cordis, Shanghai, Hongqiao, 333 Shenhong Road, Minghang District, Shanghai	上海市闵行区申虹路333号 上海虹桥康得思酒店B1层 86(21) 52639618	

chef	restaurant		
Craig Willis	**mr willis**		
p080	3f, 195 Anfu Road, near Wulumuqi Road Xuhui District, Shanghai	上海市徐汇区安福路195号3楼, 近乌鲁木齐路 86(21) 54040200	

chef	restaurant		
Steven Er	**Henkes**		
p006	Reél Shanghai Department Store 1E, 1601 Nanjing Xi Road, near Changde Road. Jingan District	上海市南京西路1601号 芮欧百货1楼1E, 近常德路 86(21) 32530889	

chef	restaurant		
Lek Chayoo	**Mi Thai**		
p093	2f, 195 Anfu Road, near Wulumuqi Road Xuhui District, Shanghai	上海市徐汇区安福路195号2楼, 近乌鲁木齐路 86(21) 54039209	

chef	restaurant		
Rick Bartam	**InterContinental Shanghai Hongqiao NECC** ｜ 国家会展中心上海虹桥洲际酒店		
p124	No. 1700 Zhuguang Road (National Exhibition & Convention Center, Gate 3, close to East Yinggang Road), Qingpu District, Shanghai	上海市青浦区诸光路1700号 (国家会展中心3号门, 近盈港东路) 86(21) 67001888	

chef	restaurant		
Kasper Elmholdt Pedersen	**Pelikan**		
p086	225 Xikang Road, near Beijing Xi Road, Jingan District, Shanghai	上海市静安区西康路225号, 近北京西路 86(21) 62667909	

chef	restaurant
Sean Clark	**The Table**
p052	3a Dundas Street EH3 6QG, Edinburgh, Scotland www.thetableedinburgh.com 0131 281 1689

chef	restaurant		
Alexander Bitterling p021	**Thought For Food** 357 Jianguo Xi Road, The Living Room by Octave, 1/F, near Taiyuan Road	上海市建国西路357号音昱听堂1楼，近太原路	
Brian Tan p024	**hoF** 30-1 Sinan Road, near Huaihai Zhong Road, Huangpu District, Shanghai	上海市黄浦区思南路30-1，近淮海中路 86(21) 60932058 微博：@hoFbar	
Eduardo Vargas P014	**Colca** Second floor, Unit 2201, 199 Hengshan Road, near Yongjia Road, Xuhui District, Shanghai	上海市徐汇区衡山路 199号2楼2201单元，近永嘉路	
Stephan Stiller p032	**Taian Table｜泰安门** 465 Zhenning Road, Lane No. 161, 101-102, Building No. 1 Changning District, Shanghai	上海市长宁区镇宁路 465弄161号1号楼101—102号	
Willmer Colmenares p108	**CHARdining｜恰餐厅** Hotel Indigo Shanghai On The Bund 29F, 585 Zhongshan Dong Er Road, Huangpu District, Shanghai	上海市黄浦区中山东二路585号 上海外滩英迪格酒店29楼 86(21) 33029995	
Edward Mair p121	**Four Seasons Hotel Shanghai｜上海四季酒店** 500 Weihai Road, Jingan District, Shanghai	上海市静安区威海路500号 86(21) 62568888 fourseasons.com/zh/shanghai/	
Scott Melvin p129	**The Commune Social｜食社** 511 Jiangning Road, near Kangding Road, Jingan District, Shanghai	上海市静安区江宁路511号近康定路 86(21) 60477638	

chef	restaurant		
Bradley Turley	**Hai by Goga**		
p002	7th floor, 1 Yueyang Lu, near Dongping Lu, Xuhui District, Shanghai	上海市徐汇区岳阳路1号7楼，近东平路 86(21) 34617893 www.gogarestaurants.com	

chef	restaurant		
Nat Alexander	**Yang Jing Bang ｜ 洋泾浜**		
p048	158 Julu Rd (Underground), Unit 5119/5121/5123, Huangpu District, Shanghai	上海市黄浦区巨鹿路158号地下一层 大同坊 5119/5121/5123 室 86(21) 53099332	

chef	restaurant		
El Willy	**El Willy**		
p071	501, 22 Zhongshan Dong Er Road, Huangpu District, Shanghai	上海市黄浦区中山东二路22号501 86(21) 54045757 www.elwillygroup.com	

chef	restaurant		
Koen Vessies	**El Willy**		
p064	501, 22 Zhongshan Dong Er Road, Huangpu District, Shanghai	上海市黄浦区中山东二路22号501 86 (21) 54045757 www.elwillygroup.com	

chef	restaurant		
Juan Gomez	**Tomatito "Sexy Tapas Bar"**		
p061	Zhong PLaza, 2F, 99 Taixing Road, near Nanjing Xi Road Jingan District, Shanghai	上海市静安区泰兴路99号2楼近南京西路 86 (21) 62598671	

chef	Brand		
William Wang	**ROUGIÉ**		
p100	For more information , please call SINODIS 86 (21) 60728700	详情垂询西诺迪斯 86 (21) 60728700	

chef	Company		
Wei Chen 陈伟	**SINODIS**		
p104	700 Wanrong Road, Daning Central Square, Building A1 South Wing, Shanghai	上海市静安区万荣路700号 大宁中心广场 A1幢南楼 86 (21) 60728700	

06

DIRECTORY OF PRODUCTS
产品名录

Good food, good living.
中国领先的食品经销商

1996年成立于上海，是中国领先的食品经销商。西诺迪斯进口和分销来自14个不同国家60余个品牌，拥有精美食品、奶制品、饼房原料、干货、糖果等2000多款产品，供应国际及本地零售连锁、网上商店、四星五星级酒店、餐厅、烘焙连锁和餐饮部门。西诺迪斯在上海、北京、广州和成都设立了四个多温度控制的配送中心。

www.sinodis.com

VALRHONA 法芙娜

创立于1922年。自创立以来，VALRHONA一直秉承它的宗旨，以传统手工巧克力制作技巧并选用全世界优质可可豆，制成品质优越的巧克力。VALRHONA 的使命亦包括与我们的专业客户建立伙伴关系，深入了解与满足他们的需要，从而帮助他们踏上成功之路。

Elle & Vire 爱乐薇

爱乐薇品牌发源于法国诺曼底，一个充满异域风情，且因其广袤的牧场以及出产高品质牛奶而享有盛名的地方。品牌诞生于Elle河与Vire河冲积形成的平原地带，其品牌名称Elle & Vire也由此而来。自从1945年品牌诞生以来，爱乐薇就一直拥有一批技艺精湛的农场主为其提供高品质的乳制品。

ROUGIÉ 露杰

自1875年在法国成立至今，已成为世界各地高档酒店、餐厅厨师，以及提供其他餐饮服务的厨师们的首选。通过与专业的食品饮料进口商和经销商的合作，露杰的美食产品已经遍布全球120多个国家和地区。露杰家族以及露杰品牌的所有员工，多年以来坚持不懈地致力于发展品牌在国际餐饮界及厨师界的影响力。

KOTÁNYI 可达怡

一切从1881年的Jonas KOTÁNYI的红辣椒开始，可达怡将全部热情投入寻找来自世界各地的珍贵香草料的过程中。时至今日，这也成为可达怡只选用高品质原料制作独特调味料的原因。可达怡凸显您在烹饪中的创造力，使烹饪充满灵感、欢乐与热情。

www.kotanyi.com

DODONI 多多尼

Mr. Cheese & Company是西诺迪斯旗下一个全新的进口奶酪平台，精选多多尼等来自世界各国的优质奶酪，为奶酪爱好者提供专业的奶酪知识、创意菜谱及搭配建议。在传播奶酪文化的同时，倡导一种健康乐活的生活理念。

www.mrcheeseandcompany.com

 精选自

KitchenAid
凯膳怡

For everything you want to make.
全能主厨，饷你所想

1919年成立于美国，致力于生产和销售家用电器，其产品包含小家电及大家电，并不断推出传承品牌DNA的简化并便于"厨房生活"的家电产品。KitchenAid凯膳怡在国际上备受好评，并以其经典独特的造型和顶尖的品质和性能，荣获多项国际大奖。尤其是明星产品厨师机，从问世以来，秉承着历经近百年的卓越质量和经典设计，成为消费者和主厨的绝对首选。

www.kitchenaid.com.cn

美国康宁餐具在中国推出了丰富的产品线,如餐具、锅具、刀具、保鲜容器及水杯水壶等,旗下品牌包括 Corelle、Visions、Revere、Corningware、Swiss Diamond、World Kitchen、Snapware、Pyrex,能很好地满足从配餐、烹饪到上桌和储存的多种需求,让现代家庭的厨房生活充满乐趣。

www.worldkitchen.cn
康宁天猫旗舰店 Tmall Flagship store : https://corning.tmall.com

Visions Flair 系列

Revere 蓝宝石系列

Swiss Diamond 瑞仕钻锅

Corningware 瑰宝系列　　WORLD KITCHEN

掌握最新资讯

Keep up to date

每周更新的食谱

With new recipes weekly on Wechat, Weibo

您可以与我随时联系
我很期待您烹饪的美食
可在您编辑的图文中 @ 我

You can connect with me anytime. I'd love to see what you cook — tag me in your photos.

- Heidi 跟我学
- Heidi 中洋生活
- Heidi 中洋生活

今天就加入
"Heidi 跟我学"
大家庭吧!